Cosmic Ordering

How To Make Your Dreams Come True

JONATHAN CAINER

Collins

Collins
An imprint of HarperCollins*Publishers*
77–85 Fulham Palace Road
London
W6 8JB

www.collins.co.uk

First published in 2006

10 09 08 07 06

4

The author asserts his moral right to be
identified as the author of this work.

A catalogue record for this book is
available from the British Library.

ISBN-10 0 00 723644 1
ISBN-13 978 0 00 723644 2

Typeset by Rowland Phototypesetting Ltd,
Bury St Edmunds, Suffolk

Printed and bound in Great Britain by
Clays Ltd, St Ives plc

Be careful what you wish for…

I believe that the very purpose of life is to be happy. From the core of our being, we desire contentment. Since we are not solely material creatures, it is a mistake to place all our hopes for happiness on external development alone. The key is to develop inner peace.

from the speech of His Holiness the Dalai Lama
at the Forum 2000 Conference in Prague,
Czech Republic, 4th September 1997

We can tune into the worlds of possibility within us; to endless levels of personal growth, to the sheer mystery and magnificence of Spirit.

Miriam Firestone, Rabbi Tirzah Firestone

We have been given a gift that is really, really precious. It's simple. It isn't going to get us anything. It isn't there to augment anything. It's there for us to be able to experience something that is already within ourselves. Simple happiness, simple joy.

Prem Rawat, Maharaji

Contents

Foreword

Once upon a time, and be sure 'twas a long time ago, there lived a poor woodman in a great forest, and every day of his life he went out to fell timber.

So one day he started out, and the good wife filled his wallet and slung his bottle on his back, that he might have meat and drink in the forest. He had marked out a huge old oak, which, thought he, would furnish many and many a good plank.

And when he was come to it, he took his axe in his hand and swung it round his head as though he were minded to fell the tree at one stroke. But he hadn't given one blow, when what should he hear but the most pitiful entreating, and there stood before him a fairy who prayed and beseeched him to spare the tree.

He was dazed, as you may fancy, with wonderment and affright, and he couldn't open his mouth to utter a word. But he found his tongue at last, and, 'Well,' said he, 'I'll e'en do as thou wishest.'

'You've done better for yourself than you know,' answered the fairy, 'and to show I'm not ungrateful, I'll grant you your next three wishes, be they what they may.'

And therewith the fairy was no more to be seen, and the woodman slung his wallet over his shoulder and his bottle at his side, and off he started home.

But the way was long, and the poor man was regularly dazed with the wonderful thing that had befallen him, and when he got home there was nothing in his noddle but the wish to sit down and rest. Maybe, too, 'twas a trick of the fairy's. Who can tell?

Anyhow, down he sat by the blazing fire, and as he sat he waxed hungry, though it was a long way off suppertime yet.

'Hasn't thou naught for supper, dame?' said he to his wife.

'Nay, not for a couple of hours yet,' said she.

'Ah!' groaned the woodman, 'I wish I'd a good link of black pudding here before me.'

No sooner had he said the word, when clatter, clatter, rustle, rustle, what should come down the chimney but a link of the finest black pudding the heart of man could wish for.

If the woodman stared, the good wife stared three times as much. 'What's all this?' says she.

Then all the morning's work came back to the woodman, and he told his tale right out, from beginning to end, and as he told it the good wife glowered and glowered, and when he had made an end of it she burst out, 'Thou bee'st but a fool, Jan, thou bee'st but a fool; and I wish the pudding were at thy nose, I do indeed.'

And before you could say Jack Robinson, there the good man sat and his nose was the longer for a noble link of black pudding.

He gave a pull, but it stuck, and she gave a pull, but it stuck, and they both pulled till they had nigh pulled the nose off, but it stuck and stuck.

'What's to be done now?' said he.

'Tisn't so very unsightly,' said she, looking hard at him.

Then the woodman saw that if he wished, he must need wish in a hurry; and wish he did, that the black pudding might come off his nose. Well! There it lay in a dish on the table, and if the good man and good wife didn't ride in a golden coach, or dress in silk and satin, why, they had at least as fine a black pudding for their supper as the heart of man could desire.

Joseph Jacobs, *More English Fairy Tales*
(New York and London: G. P. Putnam's Sons, n.d.),
pp. 107–9. First published 1894.

Variations on this tale can be found in the folklore of many lands. In some, it's the wife who gets granted the wishes; in others, the husband requests (and is granted) a rather different kind of 'sausage'. However, almost invariably, they all end with everyone going back to square one.

Despite this, our cultural heritage is rife with wish-granting witches, generous genies and lavish little people. Our cosmology involves a sky full of gods and goddesses, all bestowing various gifts and favours. Our religious traditions are similarly rich with the promise of a god who 'listens to our prayers' and who may sometimes grant our requests. And what would Christmas mean to anyone if it didn't contain a Santa Claus? It's the same all over the world. No child is born, in any country on this planet, who isn't brought up to believe in entities, deities, fairies and angels. In our neck of the woods, it's supposed to be a sign of sophistication to say you no longer accept the existence of such invisible protectors. The truth, however, is that deep down within us all, there's a part of us that still has great, childlike faith in some of those mythical spirits.

This book is dedicated to all who are brave enough to admit this … and who actually call for the help of these spirits. It is a book about wishing, a book about asking … and a book about making the impossible possible. Above all else, though, it is a book about learning to be 'careful what you wish for … '

But not so careful as to make no wish at all!

Introduction

I nearly couldn't write this book.

I wanted to. I felt I needed to. But when I actually sat down to do it, nothing came out. I spent several days staring at a computer screen, feeling ever less able to express myself. Every so often, I'd get up, go for a long walk and come back. Or I'd disappear into a quiet space and attempt a deep meditation. Nothing, though, seemed to help.

I knew I had plenty to say, but something was blocking it and I just couldn't work out what it was. It certainly wasn't a lack of personal experience. I have done quite a lot of cosmic ordering in my time.

I've successfully requested contracts, coincidences, offers, opportunities, insights and invitations. I've 'wished into my life' some wonderful people ... and some great solutions to seemingly hopeless problems. I have (or at least I seem to have) summoned traffic jams to delay meetings that I did not want to hold and, when running late for meetings that I really did want to make, I have asked for (and apparently been granted) clear roads in the middle of rush hour. As for parking spaces? Why, I've called up cloaks of invisibility to cover cars left in tow-away zones. Once, I even manifested a house . . . The need was great, the situation was urgent and, whilst I tried to remain open to the idea of another solution, it seemed that what was sorely needed was a new home for a family in crisis. I cared, and I wanted to help . . . and, for various reasons to do with availability and suitability, the only possible solution involved a purchase, not a rental. Rentals can sometimes happen very quickly. Purchases happen at the speed of a stoned snail. Usually. This one, however, was all accomplished – from finding the property, to negotiating the loan, to helping the existing occupants move out, to completing all the legal red tape, to moving in the new residents – over four straight days. It took every last ounce of my willpower *and* my wishpower. I was mentally and physically exhausted for a long time afterwards. I now realise that I put more into it than I needed to. I could (and should) have let the universe take more of the strain. But it worked. Nearly six years later, the crisis is long past and the family are still living there.

And it showed me exactly what *is* possible when you ask the universe for help and then offer to do all that you possibly can from your side, to help the universe help you.

At that time, I had no idea that I was making what some people now call a cosmic order. I was just doing what I have been instinctively doing all my life. I was trying to reach outside my limited, physical self and make myself more 'in tune' with the energy that keeps this whole world turning. I was attempting to surf the sea of serendipity; to put myself in a place where a wave of opportunity was most likely to break . . . and then wait patiently whilst remaining ready with my board, so that when the chance came I could seize it, or follow in its wake.

I didn't have a short, catchy name for this. I just used to say I was trusting that the world is, by and large, a benign and generous place . . . and that I was hoping to make it as easy as possible for the world to be that way towards me!

By then, I knew from experience that I could ask, within reason, for almost anything . . . and be in with some chance of getting it. And I had long learned, from much the same sort of experience, that it wasn't always so smart to do this.

I was definitely getting 'more enjoyable results' by being open-minded and adaptable about what ought to be happening. My more specific requests to the universe were often granted, but it seemed to me, they often had a tendency to backfire.

I had, some decades previously, become so keenly aware of this phenomenon that I became inspired to seek out an explanation for it. 'How does it all work?' I had wanted to know. What *are* these goals and objectives that people set themselves so determinedly? Why *does* the universe sometimes offer us such spookily apposite and immediate assistance and sometimes show us such apparently callous indifference? And why, even when they have what they claim to want more than anything else in the world, do so many people still feel that their lives are empty – or as if something is somehow 'missing'?

My quest for a half-convincing answer led me down many roads, literally and figuratively. I hitchhiked across America. I journeyed to the East. I studied psychology and (of course) astrology. In the process, I began to develop a lifelong fascination with eccentric philosophies. In much the same way as some people hoard books, records or china statuettes, I collected cosmic curiosities. Odd ideas. Unusual beliefs. Wacky, zany, offbeat and downright peculiar points of view.

Naturally enough, I explored magical practices and mystical insights. I researched Buddhist, Jewish, Christian, Pagan, Moslem, Hindu and Sikh beliefs. I also tried to find out as much as I could about New Age cults and traditional sects. Although I couldn't help developing my 'favourites', I tried to be eclectic. I still do.

I also tried to avoid being judgemental, although I was not and still am not wholly successful. I get along fine with any creed or culture that effectively accepts intrinsic merit in other creeds and cultures. But despite many years of consciously cultivating a wider world-view, I'm afraid I still can't tolerate intolerance!

What's more, I am still very much a traveller and not a man who has reached his destination. After over thirty years of seeking the answers to the questions I outlined above, I can happily put my hand on my heart and say, 'I still don't know.'

I am, however, enjoying the search for those answers as much as ever. And as that search has taken me up and down a lot of streets, and round in several circles, too, I have learned at least a little bit about what some people might call 'spiritual territory'. Thus, when a philosophy called cosmic ordering recently caught the attention of the British media, I immediately knew what was going on. The name was new, but the process was the same one that I had been exploring in all those ways, for all those years.

Clearly, others recognised this, too – for I suddenly found myself being invited on to prestigious TV and radio programmes to talk about this new phenomenon. Somehow, even the more sceptical, casual observers of my daily newspaper column could tell that this was something I knew about, too.

In this regard, they were quite right. Whatever's going on in the heavens, regardless of what else needs to be communicated, I am forever encouraging my readers to have faith in themselves ... and to ask the universe for help.

Nonetheless, I had a problem with that phrase 'cosmic ordering'. It is a clumsy, inelegant name for a deeply graceful process, and has the unfortunate effect of implying that our entire, amazing universe is just some kind of giant department store that does home deliveries.

I also had a problem with the way in which it was being thrust into the limelight. 'Hey, folks. Here's a great new way to get what you want. Just *tell* the cosmos what you are after! Place your order, following these simple steps . . . and all you have ever wanted can be yours.'

I felt in honour bound to point out that, while it *is* that simple in principle, it really isn't so simple in practice. Driving a car is pretty simple, too . . . but there's a very good reason why we train and test people before we let them loose on the road alone.

Technically, to make cosmic ordering work, all you need do is desire something. But if you apply this without also employing some serious discrimination, you've got a recipe for big trouble.

I happen to know that some deeply spiritual people on this planet have similar concerns. A few years back, the head of one of the oldest, most respected religions in the world banned a traditional practice involving a 'wish-fulfilling jewel'.

We live in a world where many people are so short-sighted and materialistic that they cannot see beyond their own immediate needs and wants. Under such circumstances – and until things change – techniques and methods for getting the universe to give you what you want are like sharp knives in the hands of toddlers.

It is deeply irresponsible to put such things in the hands of people without also telling them something about the potential consequences of their powers.

Fearing that such information was about to make its way into mainstream society, without the appropriate checks and balances to accompany it, I felt obliged to write something.

Yet when I sat down to tackle the topic, I found I could not say a word. Something inside me was resisting.

I tried the obvious. I asked the universe for help. I placed my cosmic order with great sincerity. But no reply was forthcoming. I just kept becoming ever more keenly aware of the many reasons why the real power of cosmic ordering may be better 'self-discovered' than taught. I kept thinking, too, about the 'unrealistic expectations' that a book on cosmic ordering might raise.

If you're dying of cancer, you can't expect a miracle cure just because you have placed a cosmic order. If you're in a part of the world that's ravaged by famine, you can't stop yourself from starving just by imploring the universe to bring you food. And if you think you can 'cosmicically order' your way to a lottery win ... forget it!

So, if cosmic ordering works, which it does, and if it produces miracles, which it can, why can't it help in all those ways?

Well, of course, it can sometimes. Some cancers do go into

spontaneous remission for no obvious reason. Some starving people do end up getting fed just when it seems all hope is lost. Some pains do suddenly stop. Some people win big money on the lottery. Usually the beneficiaries of these rare blessings will have 'asked the universe' in some way. So in their eyes, at least, there's no end to what cosmic ordering can do.

But for the rest of us ... well, it's tricky. Not everyone *can* win the lottery. On the one hand, to place a cosmic order, we must have faith in the boundless munificence of the universe. On the other ... we must not get our hopes up too high.

Religious authorities, of course, face exactly the same problem when it comes to explaining the erratic results of prayer. Sometimes, it seems, whichever god you pray to, he (or she) gives you exactly what you ask for. Sometimes, you get something different. And sometimes, it just seems as if your pleas have been rudely ignored. We're told that a fulfilled request is proof positive of the Deity – and the religion. Non-delivery is simply 'God's will'. This must be accepted in the trust that the Higher Power has a bigger plan that we mere mortals cannot even begin to imagine. 'The Lord moves in mysterious ways', or 'Everything that happens is the will of Allah', or 'God has his reasons'. Such statements are, generally, rather lacking in intellectual complexity. If they came from any other authority, they'd soon be questioned and rejected. They certainly don't come close to answering the questions:

Why *do* some people win the lottery while others don't?

Why *do* some people seemingly get all the luck while others, apparently, enjoy none?

In the supposedly more conscious, New Age fraternity, however, similar philosophies of passive acceptance hold sway. 'It's just not meant to be,' people say when they have tried their best and failed to get something they feel they badly need. Or, 'It wasn't in my destiny', or 'It must be my Karma.'

This may be fine or it may not be. All I know is that I've never heard an explanation that makes total sense to me – and I don't have one that I can pass on to you. Frankly, if I did, I'd be telling you what you ought, or ought not, to believe about the universe, under the pretext of telling you how to do cosmic ordering. Likewise, if I told you how you should or shouldn't place a cosmic order, effectively I'd be telling you how to speak to the universe. How to address it. What language it likes to hear. What gestures it prefers you to make. To all intents and purposes, I'd be telling you how to pray!

Now that's very tempting. I haven't been an astrologer for nearly thirty years without developing some pretty strong theories about how the universe works. And I've had a lifelong interest in all kinds of esoteric spiritual practices. I've read lots of book with very long words and complicated ideas in them. I've attended all sorts of lectures from all kinds of teachers, gurus and alleged experts. I've . . . well, you get the idea. So I could easily have filled this book with a mixture of what I think, what I have been taught to think, what has 'worked for me', and what I think will probably work for you.

If this book was about car maintenance, for example, or cookery, that might be fine. But cosmic ordering is about each person's individual relationship with the universe. And the moment I start interfering with your own relationship to the cosmos, the higher power, the universe, or God, I'm behaving just as arrogantly and presumptively as every priest, preacher, cult leader or self-appointed religious megalomaniac who has ever walked the earth.

Our planet is full to bursting point with ecclesiastical estate agents. Theological middlemen. Spiritual power brokers. People who purport to stand between you and the Supreme Being. 'Worship this way,' they say. 'It's the only way that works.' 'This is the true way.' 'This is what you're supposed to be doing.' 'This is what you should believe.' 'This is what God wants from you.' I don't want to add my voice to a debate that's already full of a billion voices, shouting as loud as they can.

After a week or so of this kind of struggle, I realised that for the sake of my sanity I had to let it go. I rang the publishers and told them we had to call off the whole plan. The very next morning, I woke up thinking, 'I did the right thing. I absolutely *can't* write a book about cosmic ordering.'

But then, I realised, I can't write it – but I know someone who can!

CHAPTER I

Meet your angel

Allow me to introduce myself. I am your spirit guide, your psychic servant, your invisible valet. I am your connection to the cosmos, your personal intermediary in all transactions that involve the universe. I exist only to serve one purpose: to carry out your wishes. I am here to answer your prayers and fulfil your cosmic orders.

Now, before we say another single word, I need to clear something up. I did not create you and you are not answerable to me. I have my own very deep and private relationship with the supreme Creator – as do you. Whatever you believe, you believe. Whatever you don't believe, you don't believe. It is very important that, throughout this book, I do not influence or alter, in any way, your existing view of what some people call the 'Divine'. I can carry on speaking to you only if you understand this. Are we clear? Good. Then, let's get straight to business.

You want to know how you can fill your life with more of what you want – and less of what you don't want. I am here to tell you that this is gloriously possible. And, what's more, it's easy.

What? No hidden drawback? No impossible price to pay? Absolutely none. I guarantee it. We are not talking here about some strange, archaic magic or some pact with a dark and dangerous power. We are simply talking about a natural facility. A process that is happening automatically, the whole time you are alive. A relationship that already exists, albeit unconsciously.

I am here to help you become just a little more aware of this natural phenomenon so that you can take full advantage of all it has to offer. You don't question your breath. You don't ask why it comes in and out of your body, nor do you wonder on what terms it is being granted to you. You don't wonder about how your heart beats. You don't ask, 'Why have I got a brain?' You just accept these gifts without a second thought. You use your intuition, too, whether you know it or not. Sometimes you ignore it. Sometimes you heed it. And, sometimes, you doubt its very existence.

Nevertheless, something deep within you functions like a radar screen. You get good feelings about some situations and bad ones about others. You have strong, seemingly illogical reactions to particular individuals, and though you may be able to hide this, you certainly can't deny it.

I need to point all this out to you because I want you to understand where I fit into your world. I live somewhere in between your intuition, your instinct, your intellect, your willpower, your determination, and your luck. I am none of these things ... and yet, somehow, I am all of them. I am your genie of the lamp, your wish-granting fairy, your lucky leprechaun, your magical familiar. I am the source of extra help that you have been secretly summoning ever since you were a tiny child ... and I have been waiting a long time for this chance to talk to you.

I'd like to take my time over this conversation because I want to make sure that I explain myself as clearly as possible. If you are in a mellow mood, you will hear me more distinctly. If you are just in a hurry to get to the end of this book, our dialogue will be rushed and our opportunity will be wasted. So stick with me. All should become clear if you keep on reading with an open mind. And if I repeat myself from time to time, forgive me. There are some things I would rather say too often than not enough. We are, after all, talking about something pretty important here. We are discussing my ability to make your dreams come true and your ability to aid me in that process. That, surely, is worth a little of your quality time and a bit of your best concentration. So relax, please, and give me a chance to express myself and I will endeavour to give you ... whatever you want!

First, with your permission, I had better give myself something: a name. As you will have noticed already, there are lots of names for the job that I do – and for the type of entity that I am. None of them, though, are quite right and some of them are very problematic. Nevertheless, I must give myself a label of some kind because I am talking to you ... and you live in a world where labels are very important. So, rather reluctantly, I think from this point on, you had better call me a guardian angel.

The guardian bit isn't, strictly speaking, accurate. I'm really much more a servant than a guardian. I'm a good servant, so I can be put to work as a guardian if required. But guardians should give guidance and provide protection, and for reasons that will become clear once we get to know each other a little better, we don't always do that.

If I'm not necessarily much of a guardian, I am even less of an angel. I mean, if you are talking about a disembodied spirit, an excarnate being, an esoteric, multi-dimensional, non-physical yet reasonably conscious force or presence, I guess, 'angel' is fair comment. But I'm not really an angel in the traditional religious sense. Angels are all over the Bible. They have a long complex mythology. They play harps, dress in long white gowns and have wings. They float on clouds. Sometimes, they 'fall'. Up is to down as north is to south as right is to wrong as angel is to devil. Well, you know how it goes on from here. Devil bad, angel good.

The trouble is, I can't really relate to any of that stuff. You are a human. I'm sure it makes perfect sense to you. But from where I'm standing (metaphorically speaking, of course) there are no such distinctions. Where I come from, we don't 'do' judgement. We don't do past and we don't do future, either. In my world it is always here ... and it is always now. So, if you ask me something, I don't ever sit and think, 'Hmm, do you deserve it?' I'm not Santa Claus. I don't know if you've been good or bad – and I don't care, either. I mean, I do know, in one way. I know how you are feeling, deep down inside. I know whether you think you have been good or bad and I can tell whether you are just justifying something or expressing a truly deep belief. It still doesn't make any difference to me.

I'm here to do your bidding, wherever I can, whenever I can, regardless of whether you love yourself or hate yourself. My job is to get you what you want, if it is at all possible. Ah! Thank you for noticing. That's very astute of you. You are quite correct. I did say want. I did not say need. That was deliberate. Whatever you want, I'm here to supply it. Think of me as astral room service. I'm a spiritual butler, an esoteric genie. Oh, and by the way, I'm right by your side 24/7. I don't ever go away, I only go to sleep. The nanosecond you call for me, I wake right up.

I don't have a worldly life of my own. I can communicate only in whispers, in dreams, or in quiet psychic moments. Yet, for all that, my sense of identity is quite strong. I know who I am, I know what I'm supposed to be doing – and I know who I'm supposed to be doing it for. I take all this very seriously. It gives me great pleasure. I feel fulfilled when I am carrying out my role, and the rest of the time I feel nothing at all.

When I'm not being called upon, I become a kind of silent witness. I just sit around, passively observing the soap opera of your life. I don't look back, I don't look forward, I don't go off and play Scrabble with the other guardian angels, I just watch you on a kind of TV screen until you remember that I am available ... and that you can reach out for me. Then, I do my best to fulfil whatever request you make; after which, as long as I have been reasonably successful, I just resume my position in the audience.

Infinite possibilities

Is there a limit to the number of things you can ask me for? You must be joking! Do you know why you think there has got to be a limit? Because you are a human, a mortal. Once upon a time, you didn't exist. Not too long from now, you won't exist again. Or, at least, not in the form you are currently taking. Something within you is infinite, but your shape, your frame, your very identity, is distinctly, definitely and sometimes rather depressingly finite.

It isn't that way for me. I have been here forever. I'm going to stay here forever and though my relationship with you is going to change when you move on, nothing else is going to alter much.

Here in the midst of eternity and infinity there are no limits, no restrictions. Three wishes, three thousand wishes, three million billion wishes ... it's all the same to me. Really, you should feel free to use me more often. I'm not like a car that's going to need its engine changing if you drive it too far. I'll try my best to do anything for you. Anything you want. Any time you want me to. As often as you like. And I can promise you this: no matter how often in your life you have already taken advantage of my services, you could have done it a lot more. I really would not have minded. I would have been glad to help. Nor do I have any problem with the fact that you didn't ask me as much as you might have done. I live in the here and now, remember. I don't do judgement and I don't do resentment, either.

Under the strict cosmic laws that govern our relationship, I cannot tell you how you should be living your life. I can, however, drop a few hints and hope you catch them.

You might start by wondering what I am doing in your world. Does every human being have a guardian angel? Absolutely. There is exactly the same number of us as there is of you. Do we communicate with each other? As much as we need to. Do we sometimes find ourselves in conflict? Ah. Now there's a good question.

What happens if one human requests the Koh-I-Noor diamond and then another human asks for the very same thing? Do their respective guardian angels go off and have a wrestling match to see which one can win the honour of taking back the prize? Not quite. We simply don't fulfil orders of this kind. We can't. It's much better if you are less specific. If you just ask for the Koh-I-Noor, you're asking for something that belongs to someone else. Not only is that someone else asking their own guardian angel to help them keep it, they are employing a lot of earthly guardians to make sure it stays in their possession.

I am going to back away the moment you come into conflict with someone else whose guardian angel is on an identical mission to me. It doesn't matter if one of you is an honest, law-abiding citizen, and the other is a ruthless, merciless fiend. It's not as if the two guardian angels have to sit down and have a conversation. 'Okay, your candidate is clearly a nicer person than mine, so I will step out of the way.' Niceness doesn't come into this, because judgement doesn't come into this.

What does come into this is a kind of polarisation. The two guardian angels effectively cancel each other out. They exert an equal and opposite force on each other. They remain locked in a kind of stasis for as long as the two human beings in their care are both striving for something that only one of them can have.

This is why, whenever there is a situation that contains plenty of conflict, you rarely see much evidence of a guardian angel at work on either side of the argument. It is only when one person finds the wisdom to back down, let go and reach, somehow, for resignation and acceptance, that both guardian angels are effectively released from the state of perpetual stalemate.

So, if you want your guardian angel to work well for you, you need to make requests that the angel has a chance of fulfilling, keep your angel away from 'black holes' like the one I have just described, and work within the limits of what your angel can or cannot do.

Ah! Well spotted. That was very astute of you. I did say, several pages ago, that I existed in a world that was limitless. That's true. I do. I have the whole of eternity to work with, so if I get caught up in a conflict, it is really no problem to me if it takes a few hundred years to resolve itself.

But while I can wait forever, you can't.

Actually, that reminds me of a joke. I'll share it with you now.

* * * * * * * * * * * * * * * * * * * *

There is this person who is desperate to speak to God. They pray and they pray – and they badger their guardian angel for some assistance. Eventually God decides to grant them an interview. The supreme Creator manifests in the form most appropriate to this person's religion and proceeds to say in a deep, booming voice, 'What do you want to know, my child?'

Our deeply humbled and awestruck individual gathers their composure and, after making various gestures of gratitude and respect, says, 'I really want to understand the meaning of time. Could you help me to comprehend something of its true nature?'

The temporary incarnation of ultimate, infinite consciousness replies, in a slightly kinder and rather less booming voice, 'Okay, try to imagine the passage of a million years. That, to me, is but a second in all of eternity.'

The person stops for a while to consider this mind-boggling explanation. Human beings have not been alive on the earth for a million years. Even the Sphinx, in ancient Egypt, is no more than a few thousand years old. Though it is possible to imagine many civilisations that may have come and gone before this, leaving no trace, it is clear that the entire journey of human evolution has taken place in the blink of an eye. One single life, no matter how long it lasts or how much is achieved within it, is entirely insignificant by comparison. A million years is no more than a single second. It's unimaginable.

Suitably humbled, our seeker of spiritual insight summons the courage to attempt a supplementary question. 'Dear God, what, then, is the real meaning of this stuff called money ... which we human beings seem to value so dearly?'

The Creator, by now, is speaking in a much softer tone and adopting an almost chummy demeanour. 'Well, try thinking of it this way. From the perspective of supreme spiritual wisdom and knowledge, a million pounds has no more or less meaning or value than a single penny.'

Not wanting to waste the opportunity of a lifetime, our questioner then attempts a third inquiry. 'In that case, could I have a million pounds?'

'Sure,' says the Creator. 'Just wait a second!'

* * * * * * * * * * * * * * * * * * * *

CHAPTER 2

What can I do for you?

Now, perhaps you are starting to feel disappointed. After all, it must be clear that if even God can't guarantee you an instant million, your guardian angel stands no chance of organising such a development. But this lottery-winning thing is tricky.

You want to win the lottery. So does every other person who has purchased a ticket. You have asked for my help in this. They, too, have beseeched their guardian angels to do the same. So what do you think happens in the invisible realm, when those numbers come rolling out of the machine? Hundreds and thousands of guardian angels find themselves caught up in that black hole of cosmic conflict. They all have the power to influence the outcome – their power is equal, but they are all trying to influence it in different ways. The upshot is a kind of mass paralysis.

You are right, though. Somebody still has to win it. So, what then, if not a guardian angel, determines who that somebody is? In all honesty, I'm still trying to work it out. It's one of the things I ponder over while I'm sleeping, in between carrying out your requests. What I can tell you is this. Play the lottery if you want to. Don't if you don't. If, though, you are looking for ways to engage the help of your guardian angel and improve your lot in life, try understanding more about how your angel is best able to help you.

Let's go back to that Koh-I-Noor diamond for a moment. If, instead of requesting the Koh-I-Noor itself, you ask for 'a diamond that's rather like the Koh-I-Noor', you slightly increase your chance of getting one because you are placing less of a limit on what your guardian angel has to do for you.

If you can be even more flexible and request 'the wherewithal to attain a diamond', you are definitely helping me to help you. I will always do what I can on your behalf, but I need you to put in some effort, too.

So, even better than, 'Hey, guardian angel, please fetch me what I require,' is, 'Hey, guardian angel ... I've got a plan that might gradually put me in a position where I could afford a nice big diamond.

Could you work with me to ensure that each step I take, along the road to carrying out this plan, is as smooth and successful as possible?'

Now, that's the kind of question I like to be asked. I can, indeed, help you out there.

Nevertheless, I can only ever get you what's attainable and if it can't be had, I can't get it for you. If there is only a remote chance of it happening, I have a very remote chance of helping to make it happen. And, of course, if it is relatively easy to get, it's no problem for me to come through with the goods for you. That probably sounds blindingly obvious, but if you really think about it, it is not so simple. From my angelic perspective, I can often identify opportunities to which you are oblivious. Often, too, I can see that some of the things you ask me for are things that you are entirely able to attain on your own.

Imagine a scale – a sort of temperature gauge with numbers climbing higher as you go along. Down at one end, you've got cornflakes. Up at the other, you've got first prize in a multimillion-pound lottery. You really don't need my help to get you breakfast, although you are perfectly welcome to ask for it, of course – I have already explained that I never object to being called upon. But unless you have just spent a night out in the wilderness, miles from anywhere, and you have woken up to realise that there is no way of finding any food, you don't want to sit around trying to summon me, when you can just walk over to the cupboard!

We've already dealt with the lottery so we won't go there again, but somewhere in between those two extremes are some other highly desirable but extremely unlikely aspirations and wishes. The question is how unlikely are they? The answer is, more often that not, they are more feasible than you might think. As it happens, I'm very good at reaching for things that seem impossible to you, but which are actually very easy, once you know how.

You might think, for example, 'Oh, I'll never get the really great job that I've been dreaming of.' Or, 'I'd love to live in a particular kind of house.' You can look around and see no way to make this happen. But I've got an advantage over you. I am excarnate. I don't live in a body. That makes me a bit like one of those superheroes in the comic books. I can zoom around the whole world in the time it takes to flick a light switch. I can see through walls, I can read minds, I can sense, instinctively, where hidden secrets are to be found. So I can go out looking on your behalf. Very often, I can find exactly what you want and see a clear way to help you get it. Then, all I have to do is wait until you ask me for my guidance.

Ahem ... Thankfully, I'm blessed with great patience.

There are times when, if I wasn't a cosmic being with deep wisdom and infinite understanding, I could get incredibly frustrated. You call me up, you tell me what you want, I line up the whole thing beautifully for you, and then we run into an all-too-familiar problem.

No matter how many hints I try to drop, you just won't hear them. No matter how many signs and signals I send you, you pay them no heed. You cook up some scheme of your own that you think is absolutely brilliant and then you think, 'That's fine. I don't need my guardian angel's help any more. I have seen my solution and all I have to do is apply it.' Meanwhile, I'm sitting and watching. I can see the flaw in your plan. I can't tell you about it, though, because you are no longer in 'angel consultation mode'. So I just have to wait until it all goes horribly wrong and you finally come back to me and say, 'Help!' once more. Then, I have to hope that whatever I had lined up is still feasible.

Usually, it's too late by then and so I have to set up a brand new opportunity. I don't mind this, of course, because I'm highly enlightened and endlessly sanguine. I don't even mind it when you blame me for the fact that things haven't worked out as you hoped they would. When you ask questions such as, 'Why did you lead me down such a crazy road?' I don't feel the need to reply, 'I didn't lead you there. You set off without me and did the whole thing by yourself.' Really, it's no problem to me. I'm your guardian angel. I'm here to serve. Always here.

all you have to do is become a bit more receptive

Sometimes, though, if I am honest, I wish you'd talk to me more often. Even more crucially, I wish you'd trust me and listen to me. We could achieve so much together, you and I, if only we had a slightly deeper, more meaningful relationship. I could show you so much. I could work with you so successfully. We could make a brilliant team. You want more money? I could show you how to make it. You want less stress? Boy, could I be useful to you. You would like your love life to be more fulfilling? I could give you so many helpful clues and pointers. You would like the right things to happen in the right way at the right time? I could steer you in the direction of some truly tremendous coincidences. All you have to do is become a bit more receptive.

I'm not suggesting this then allows you to become lazy. You would still have to put in plenty of effort and energy, but it would tend to pay off very efficiently. Between us we could increase the miracle quotient in your life by a significant amount. We could do it by spotting possibilities that seem miraculous, but which really just require a little faith, imagination and commitment.

Dungeons and dragons

Now please, don't get me wrong here. You wouldn't end up leading a fairy-tale life. Or rather you would, but, in the manner of every classic fairy tale, it would have dragons to conquer and villains to contend with.

No guardian angel can entirely protect you from these. The best I can do is to help you to handle them better. Even when you are down at the bottom of some deep, dark hole, I can find ways to help you get out. I can, for example, work with your deepest survival instincts, and as long as something inside you is listening to me, I can encourage you to shout out 'Help!' just at the moment when someone happens to be walking past the top of that hole with a very long ladder in their hands.

It is possible, though, that my solution may not always be so simple. Neither of us can have a worthwhile relationship if we're not truthful with one another. So, at the risk that you'll be put off by my candour, let me take you back to the bottom of that hole.

You have fallen into it, and now you are stuck there. Maybe this was because you weren't listening closely enough to me or maybe it was because, no matter how I tried to steer you away from it, I was up against some other cosmic force that I couldn't get the better of. (Remember, I may be powerful but I'm not 'all-powerful'.)

Anyway, here we are down this hole carefully avoiding an argument about how we have got there or whose fault it is. Unfortunately, there is no friendly passer-by with a ladder. The best solution I can see for you is a network of underground caves and tunnels that you can somehow dig your way towards. Eventually, they lead to daylight once more ... but they wind on and on for many a mile before they do this.

I can identify the route and I can communicate with your intuition clearly enough to get you digging. There is, though, no getting away from the fact that the journey is going to be stressful, exhausting, uncomfortable and claustrophobic.

I'm going to say, 'Keep on crawling.'

You are going to say, 'Why? So we can get even further into a highly unpleasant situation?'

I'm going to say, 'Yes, because it's the only way out.'

You are going to have to trust me. And then, you are going to have to continue trusting me even though there is no sign of an improvement in your circumstances.

This kind of thing isn't going to happen very often, but it is going to happen in the future and it has happened in the past. It's part of the reason why now you are slightly sceptical about my very existence. Truthfully, the fact is that even with me right by your side, there are going to be times when you find life tough. As your guardian angel, I don't just help you to enjoy an endless succession of good times – I also have to do my best during the bad times that neither of us can entirely avoid.

I don't just help you to enjoy an endless succession of good times

I'm telling you all this because I want us to have a realistic relationship. If we are going to work well together, we have to be partners. Whilst partners must always respect each other, they are not going to get very far if one puts the other on a pedestal.

So, let me be clear. Yes, I can work wonders for your love life, and yes, I can help you become materially better off. But there will always be some things that

I cannot do for you in quite the way you might wish. And there will even be times when your particular circumstances make it hard for me to do even the simplest things for you.

Let me take you back to that sliding scale of attainability, the one with cornflakes on one end and lottery wins at the other. I hope I didn't leave you with the impression that cornflakes were more trivial than lottery wins. Under some circumstances, breakfast can be a very difficult thing to get. And if you happen to live, for example, in certain parts of Africa or India, the same is true of lunch and supper. What if you are starving? What if all you want in the whole wide world is food? Suppose you are in a village, ravaged by famine, miles from the nearest international aid centre? What if, right where you are, food is up there, on the far end of the scale, next to a major lottery win?

> I am powerful but not 'all-powerful'

I am still going to be with you, doing my utmost to help steer you in the direction of sustenance, but, as I explained before, I am powerful but not 'all-powerful'. Here, we're in a different situation. You're not asking me for a new car. If, for some reason, I can't help you get such a thing, it's not the end of the world.

Besides which, I can always supply you with something that's potentially just as valuable, if not more so. I can help you, if you ask me, to feel fine about the fact that you haven't got your car. I can help you overcome your desire for it. I can help you decide that life is so good, just the way it is – and that you are so blessed in so many other ways, that the car was really just a responsibility that you didn't need.

But is that really the same if you are starving? Or if you are losing someone who means the world to you? Or if all you want, more than anything, is to bring a loved one back to life?

I can supply you with something just as valuable

Actually, yes, it's exactly the same. There is still something I can do for you. I can help to wrap you in a bubble of faith and serenity. I can help you give in gracefully. I can help you to accept and to somehow see the wonderful, even in the seemingly terrible. Even if you are terminally ill and there is absolutely no way to find a cure, I can at least help you cross from this world to the next in a state of great love and trust. And, one day, no matter where you are or what kind of life you have led, that is, indeed, what I will be doing for you.

Still, we're not there yet. And before we get there, we've got a lot of other things to get ... and to get through.

CHAPTER 3

How should you talk to me?

When there is something that you want from the universe, what's the best way to let it know? If you are trying to summon my assistance, how do you get it?

Are there special rituals that you need to perform? Or profound states of mind that you need to enter? Do you have to light candles, burn incense, wear special clothes, or sit facing a particular direction? Are there specific incantations that you need to recite at particular times of the month? Is it better to seek me in the morning or at night? Should I be shouted at or whispered to, pleaded with, or reasoned with? Do you need to make a list and put it under your pillow? Will our communication be better if you go out into the wilds of nature or sit in a quiet cosy space in a still state of meditation?

These, and many more such suggestions, have been made by various 'authorities' and it is not my place to agree or disagree with them. From where I stand (or, to be more accurate, float – for we must always remember that I have no physical form), it's very simple. What's sacred to you is sacred to me. What makes you feel good makes me feel good, too. Whatever you believe, I can accept. There is no right way or wrong way. There is only a way that is heartfelt – or a way that is insincere.

I tend not to listen to you when you are being insincere. This is not because I disapprove of insincerity in any way (as you know, I cannot judge, because I exist in a realm where there is no judgement). I can certainly see that insincerity is a fine thing in its own way. The world would be a very dull place if human beings meant everything they said from the bottom of their hearts. There would be no jokes, for one thing. There would also be a lot fewer adventures. In a totally sincere world, everyone and everything would have to be totally straight. Nobody would be able to experiment or explore, far less posture and pose. Acting would go right out of the window and, with it, storytelling, drama, theatre and cinema. There would be no fashion, no fun, no froth ... to say nothing of wit and whimsy. Artists would have no need to explore subtlety and nuance since everything would be obvious to everyone. Without being free to experiment with differing levels of inner commitment, life would lose a lot of its colour, its mystery and its charm.

I exist in a realm where there is nothing but sincerity, which is why I appreciate the existence of insincerity in your realm. So, when I say that I don't tend to listen to you when you are being insincere, I simply mean that I need you to be sincere when you speak to *me*, because if you are not, you are going to end up asking for a bunch of things that you'll rapidly regret. An inbuilt safety mechanism, in the bridge between our respective realms, ensures that nothing can ever pass between the two of us unless it is truly, deeply, profoundly 'meant'.

So, how do I tell whether you mean something or not? That's easy. I just look at you. You may not always be completely sure of the level of your own sincerity, but to an impartial observer like me it's an easy thing to recognise. I can see into your heart. I can tell what the deepest part of you is feeling and wishing for. Usually, though not always, when you really mean something, you will do and say certain things. There will be gestures that you make, actions that you undertake, stances that you adopt, and processes that you go through.

Usually, too, there will be ceremonies you perform that somehow help you to get into an appropriate mood of clarity and intensity. These ceremonies have no meaning to me ... other than in the meaning that they have to you. Really. It makes no difference whether you go down on one knee, whether you clutch a cross, carry a crescent, wear a six-pointed star, or draw a circle in chalk on the floor. You can make your request using long or short words in any language you like ... in any tone of voice that seems right to you. You can be alone or in company, at home or in a strange and distant place. If it turns out that the way you need to express yourself, from your heart of hearts, is by putting on a pirate's hat and a bathing costume and then pedalling on a unicycle whilst whistling 'Waltzing Matilda' and juggling soft-boiled eggs, I'm not going to have a problem with it.

Conversely, you can dress from head to toe in your most sombre, serious clothes and perform all the actions that you have been taught, from an early age, to associate with 'respect for a higher authority', but if, secretly, a part of you still doesn't really mean it, it's going to be very clear to me. (Not, by the way, that I *am* a higher authority. Here in my realm we don't do comparisons, any more than we do judgements. We leave all that to the humans. Round here, there is no up and no down. No higher or lower and no bigger or smaller. There is no 'important and unimportant'. There is only the matter of feasibility.)

Don't feel that you ought only ever to come to me with what you consider are 'big issues'. It makes no difference from my point of view whether you are asking for a parking space or a palace. If I can help, I will.

Nor should you feel that, like so many of those fairies in the storybooks, I have only a limited number of wishes that I can grant you. I don't keep a count. I haven't got a clock. I don't say, 'Right. That's seventy-two times today that you have disturbed me with requests for silly little favours. You have used up your supply and I am shutting up shop, so don't bother coming back to me with another request, regardless of what it may be for.'

You are the one who lives in the finite world. Those limitations tax your mind, not mine. I'm just here to help. Always. In my astral realm, we don't do size and scale any more than we do hierarchy or meritocracy. We just measure two things: sincerity and gratitude.

I haven't mentioned gratitude before. I find it slightly difficult to talk about because it's not the sort of thing that anyone, even a guardian angel, ever ought to ask for. Gratitude has to be spontaneous and it has to be real. If it comes attached to the slightest thread of obligation, it is somehow impure and artificial. You can say 'thank you' a thousand times and not mean it once. Or you can never say it, but you can feel it – and that counts for more. I certainly don't want gratitude for my help. You can feel free to take me for granted. It will never stop me from doing my level best to assist you.

Something inside you, however, will grow cold and insensitive if you don't stop long enough to allow a deep sense of appreciation to well up in your heart from time to time. This, in turn, will ultimately reduce your ability to be sincere – and thus to make requests that I can respond to.

Think back to when you were little and your parents taught you to say 'please' and 'thank you'. In a strange way, these three little words, which mean so much in polite, earthly society, mean even more in the astral realm. If a request comes in with a strong enough, deep enough 'please' attached to it, a guardian angel cannot help but pay attention. The 'thank you', though, really ought not to be linked to a successful delivery of your cosmic order. It ought simply to be a 'thank you for trying on my behalf'.

When you are talking to me, you should put in your 'thank you' at the same time as your 'please'. If you can manage to keep the 'thank you' going, regardless of the outcome of my endeavour, things will be more satisfactory for both of us. Your very willingness to experience gratitude will ensure that if I supply you with the best possible solution, you will appreciate it even if it doesn't match your expectation. That will allow me maximum creative flexibility in my effort to assist you ... and will ensure that you always get the very best from me.

put in your 'thank you' at the same time as your 'please'

When should you talk to me?

I am your guardian angel and I am constantly by your side, listening out for a sincere request from your heart of hearts. I don't ever say, 'Oh, but you are asking me on a Saturday and I always take Saturdays off.' I don't look to see whether the Moon is new or full, nor am I more amenable to an approach whilst

Venus is aligned with Mars.

You, on the other hand, may well find that under particular cosmic conditions it is easier or harder to get in touch with your own true feelings. New and full Moons govern the tides and they also seem to have an impact on the tides of emotion that ebb and flow within the hearts of humans. More people make their requests at these times of the month than at other times, but guardian angels are ready for duty at all times.

Now, how are you doing with all this so far? Is it making sense to you or are you feeling slightly confronted by some of what I have been saying? Perhaps you are of the opinion that I am stating the obvious and labouring the point? Yet what I have to say is crucial, and not always quite as simple as it sounds.

I have, for example, just gone to a lot of trouble to tell you about the importance of 'really meaning' a request. I have stressed sincerity, yet, if you think back, you can surely recall moments in your life when you wanted something from the absolute bottom of your heart, yet you never got it.

What was the problem there, then? Was it up at the impossible end of the attainability scale? Not really. Was it something that directly conflicted with someone else's desire and thus put me into a 'frozen opposition', or cosmic conflict, with someone else's guardian angel? Not necessarily. Were you asking me in the wrong way? How can you have been? We have just established that there is no right or wrong way to ask me.

So why didn't I deliver? Well, I tried. But once again there was a problem, to do with your level of sincerity. But this time, it was not that you didn't want it. It was that you wanted it too badly. You cared too much. You were, not to put too fine a point on it, pretty desperate.

Now normally, I can steer you towards an opportunity or fix it so that something useful crosses your path. For this to work, however, you have to be keen, alert and trusting ... but not overwhelmed with a hopeless sense of need.

It wasn't that you didn't want it. It was that you wanted it too badly

One side effect of you wanting something too much was that you weren't just asking *me*. You were making various other impassioned approaches in your immediate environment. People who are overly anxious create a kind of psychic force field around themselves. Like magnets, they try extremely hard to attract certain things and, just as magnets can repel each other under certain circumstances, they can end up keeping at bay the very thing they are so hungry to have.

This is especially the case whenever success depends on the co-operation of another individual. If someone turns to us with a smile on their face and says, 'I wonder if you could help me make this happen?', folk are normally happy enough to assist. A little bit of the guardian angel spirit exists in us all. When, however, they see that someone is aching, yearning and burning to reach a particular destination or fulfil a wild desire, they become more inclined to make their excuses and leave. They find that desperation deeply unappealing. It alienates them.

Now, as a guardian angel, I don't particularly mind it. It is not for me to judge whether your level of sincerity is inappropriately high. Your friends, your loved ones, your colleagues and companions will not be so reserved in their judgements. As it's often useful to have their help, understanding and co-operation as well as mine, you may well want to work on cultivating a degree of detachment. It will make you feel better about many things, too!

Often, when you ask me for something in a tone of manic intensity, I register your desire and do all that I can to make it happen for you. But because you want it too much, you end up repelling it so that it cannot come about. Then, eventually, you 'give up'. You let it go. You decide that you are not bothered any more. The moment you truly feel this way, all that you once wanted so badly comes rushing into your world like a river that has burst its banks.

then, eventually, you 'give up', you let it go

The irony is that, by then, you really don't want it any more. So, you are left with a difficult choice. You either have to turn away the very thing that you wanted so much, or find some way to rekindle the spark of desire. That isn't always as easy as you may think. Some people end up feeling obliged to accept the late delivery of their request and then spend many years trying to persuade themselves that it is still what they want, even though they have actually outgrown the desire.

Dawn and twilight

There is another factor that can come into play here, too. No matter how many chances I put before you, you have to seize them and make them work. That requires a degree of level-headedness on your part, but passion and perspective are like sunshine and starlight – you rarely see both at the same time. As soon as you have a lot of one, the other vanishes.

For you and I to work well together, I need you to be able to reach a state of psychological twilight. At dusk each day, there is a fleeting moment when the Sun has just started to set below the horizon, but its rays are still illuminating the sky. Then, briefly, you can see the stars at the same time. A similar phenomenon occurs at dawn. In that brief moment, it is possible to straddle two worlds.

Likewise, it is possible to lower the flame of your inner fire without extinguishing it completely – so that its heat does not repel someone who may be approaching it with a pile of potential fuel. Or, to put it all another way, you need to find a way to care and not care, all at the same time. To be detached, yet engaged; involved, yet aloof. The most easily fulfilled cosmic orders are generally the ones that are made in this spirit.

Tiger taming

So, if you do find yourself in such an uncontrollable condition, what else can you do? What if you have an obsession that just won't go away? Or an almost painful ever-present urge to bring something (or someone) into your life at any cost, under any circumstance?

Well, you could try asking *me* for help! Instead of saying to me, 'Please give me this', you could say, 'Please give me the ability to see straight.' Or, 'Please help me feel a bit less overwhelmed by my desire.' You don't have to ask me to take it away entirely. You could simply acknowledge that it seems to be getting out of control and there doesn't seem to be much you can do about this.

I have to tell you that no matter how hard it is for you to tame your own tiger, for me, on our scale of attainability, it is cornflakes! The moment I hear you asking for this particular kind of help, I can zoom into action faster than a fire engine. I can calm you down, chill you out and restore your sense of sanity – as long as that's what you want me to do. All that's needed on your part is a reminder call to me, each time you find the heat starting to rise once more.

Up here in the realm of guardian angels we do listen to music occasionally. Top of the guardian angel hit parade is that song by Sting, 'If You Love Somebody, Set Them Free'. You may think of it as a golden oldie, but round here it is still a chart-topper. Indeed, before he wrote it, I can hardly remember what we guardian angels used to sing to people to convey that same message.

These days, most people understand the principle implied in the song's title. In theory, at least, they get it. Then, however, they have a dangerous idea. They think, 'Aha, so, if I want to get this, all I have to do is stop wanting it.' Then, they try very hard to stop wanting it just because they figure that this will help them get it.

It doesn't work. We guardian angels can't cope with complicated requests of this nature. You have to really, sincerely, want to let go. Then you have to be prepared to take your chances – and to make your choices.

You can, though, always try the following request: 'Dear guardian angel, please help me not to want this thing that is tearing my heart apart and please make it easier for me by helping me get excited, instead, about something I actually can have.' If I were you, that's definitely what I would be asking me for.

Let's avoid misunderstandings

I have heard some people say that you should only ever ask the universe for help once. From my point of view, repetition is no problem. I certainly won't get offended if you keep reiterating your request. Nevertheless, it's true that you only *need* to ask once. It's possibly also true that repeating a request to the universe for help can have a strange effect on your frame of mind.

* * * * * * * * * * * * * * * * * * * *

Once upon a time, there was a woman who lived in a house on a flood plain. One day a terrible storm blew up and the waters rose as never before. Eventually, she had to take refuge on the roof of her house. However, she had great faith and so she simply sat there, praying to God. 'Dear God, I have great faith in you. I trust that you can make these waters go back down.'

She kept on praying, ever more fervently, and, despite her physical discomfort, she found she was actually enjoying the sense of spiritual attunement. After a while she was interrupted in her reverie by a lot of raucous shouting. She opened her eyes to see some people on a boat below her. 'Climb down and get on board,' they said. 'We'll get you out of here.' She waved them away, confident that the Creator would come to save her.

The water kept on rising, the woman kept on praying and another boat came by. Once again, she waved it away and continued with her intense private process of supplication.

By now, she was in a state of high elation and was feeling very sure that her gesture of faith in the Almighty was sure to be rewarded. When yet another noisy boat came by she dismissed it with hardly a glance. Still, the waters kept on rising until they covered the whole house and carried her away.

Immediately after she drowned, she went to heaven, where she urgently demanded an audience with God. 'Hey,' she said, 'What's the big idea? I prayed with all my might. I showed you all the devotion I could muster and yet you let me die?' God looked at her in disbelief. 'For crying out loud,' He said, 'How can you say that? I sent you three boats!'

* * * * * * * * * * * * * * * * * * * *

Whenever we guardian angels tell this joke it always gets a laugh – followed, usually, by a groan of recognition. We have all known people like this woman. It's amazing how many of us have a tale to tell involving misplaced expectations or messages that have been misunderstood. Mainly, they stem from overly specific requests.

Overly specific requests are the bane of every guardian angel. You'd be amazed at what people ask us for. We get orders that contain dimensions down to the last millimetre, or colours that have to be checked against the official Pantone register. Sometimes we can oblige. Usually, when we do, we can't help wondering why you felt the need to be quite so particular. I mean, sure, if you are looking for a rare widget from a 1975 model to finish restoring your car to its former glory, by all means ask your guardian angel to help you in the search. When, though, you ask us for a partner who is exactly 5ft 7in tall, with strawberry blonde hair and a degree in biochemistry, we can't help wondering whether it might be wiser to prioritise a few other qualities – such as, for example, kindness or sensitivity or a compatible sense of humour.

The trouble is, many people think they know exactly what they want. This often seems to stop them from finding out what it is that they *really* want. I shall return to this theme later. Next, though, while that joke about the woman on the roof is fresh in your mind, I am going to talk to you about signs and signals. We have spoken about the way that you can communicate with me. I have done my best to make it clear that I can hear you, no matter what tone of voice you take – or indeed, what language you speak. You don't need to go anywhere special or do anything significant to get my attention. You just have to be sincere about your request and, somehow, I will hear it.

Sadly, it doesn't work in the same way the other way round. I sometimes have the darndest difficulty getting through to you. I try rearranging your circumstances, in the hope that these will send you a hint, and you blithely ignore the message that ought to be so obvious. I'm sometimes tempted to send you an anonymous letter or flash a brief message on your TV screen. But then I remember, most of what people hear on TV goes in one ear and out of the other ... just as most of the dreams that people have are soon forgotten. But there is a way ...

CHAPTER 4

Signs and signals

Every single night, while you sleep, you hold a vivid and deeply engaging conversation with your unconscious mind. You get presented with a pageant, a colourful cavalcade of images and ideas. You become deeply transfixed by these, no matter how weird or wonderful they seem. You accept them without question and become so engrossed in them that you forget yourself.

Often, I will try to reach you at such times. It is relatively easy for me to slip into your dreams and play an influential part in them. The trouble is, the moment you wake up you tend to forget them completely. If you remember them at all, you hold on only to snippets. Even these, you feel somewhat dismissive or incredulous about.

To make them influence you, I need to make your dreams truly memorable, but I am loath to do this because it is a trick that is also often employed by your inner saboteur: your fear, your doubt and your anxiety. It, too, likes to visit you when you are asleep and it is adept at the art of creating powerful nocturnal visions. It knows how to scare you stupid by creating entire tapestries of worst-case scenarios and weaving these into a veritable panorama of paranoia.

Now, given half the chance, I can also paint an unforgettable picture. I can make you a dream that simply oozes ecstatic inspiration. I can fill your inner-cinema screen with dramatic images of hope and happiness. I can leave you feeling, when you wake up, as if you have been visited by an angel, which, indeed, you have! However, I can do this only if you leave open a small chink in your psychological defence mechanism.

If you are hanging on tightly to a resentment, or you are deeply disturbed by a worry, it is exceedingly difficult for me to reach out to you – even when you are unconscious. The dark-dream weaver will always get there before me, twisting and distorting those causes for concern like some manic puppeteer, pulling on the strings of a set of grinning, dancing skeletons.

If you want me to speak to you in your dreams, you need to invite me into them. It doesn't take much, just a brief request at bedtime for a little inspiration to reach you in the night. Keep on asking for this and, even if I can't get through that very same evening, I'll sooner or later give you a reason to smile in your sleep.

In the absence of such an invitation, I must somehow find a way to visit you during the day. My favourite way to contact you is still through your own subconscious. I like it best when I can speak to you as an 'inner voice', and you can hear and respect me. The trouble is, this too doesn't always work.

For one thing, there is a problem with the volume. When I shout at the top of my voice, you hear it as the gentle rustle of a passing breeze. And even if you're good at tuning in to subtle signals, there is another problem caused by my need to broadcast on your inner frequency. I have to get in the queue alongside a bunch of other voices with pressing messages to communicate. Depending on the time of day – or the month – there can be a lot of these.

Your ears are trying to tell you everything that they can hear from the moment the alarm clock assails them as it rouses you from your slumber. They will report each strange, suspicious sound, each new piece of audible intelligence as it arises, and demand that you give it immediate evaluation. Or, at least, that's what they will do if you haven't got the radio or the TV playing. Then, they will just be relaying to you every drumbeat, guitar chord, or violin motif; every nuance of the broadcaster's voice; every detail of the news and weather. It's hard to get a word in edgeways over all that, yet it is by no means the only obstacle I must overcome.

Your eyes, meanwhile, are allowing in an absolute ocean of information. Even while they are still blurry from a night of rest, they are processing colour, texture and quality of light. Once they start to focus on your morning paper, I've got no chance of talking to you. Suddenly, your mind is racing all over the world. You are thinking about people you have never met, politicians you approve or disapprove of, celebrities you have an interest in, and situations you think you want to learn more about. You are glancing at adverts, gazing at pictures, and studying the stories of the day that stir up your emotions ... all the time forming opinions about them at the speed of light.

Your senses of touch, taste and smell will also be adding to this deluge of distraction. All day long, ceaselessly and tirelessly, the ambassadors of your five senses will be jockeying for position in their attempt to gain the full attention of your central self. And in your day-to-day physical existence, your stomach is so loud and domineering that it can practically never be ignored!

Like someone born to hold an office of high command, you take this constant bombardment of instruction and information almost effortlessly in your stride. Yet it all absorbs more of your energy than you realise, and it means that if I want to send you a sign or signal, I'm going to have to come up with something pretty impressive. Either I am going to need to engineer some unexpected event that stops you right in your tracks, or join all the other voices in the queue for your attention.

There are, you see, plenty of other wraithlike beings that inhabit your inner world. Imagine yourself as a grand ruler on your throne. There is the head of your Organisation and Planning department, forever waving diaries, calendars and to-do lists in front of your nose. There is your social secretary, constantly reminding you to think about the people who matter most in your world and the commitments that you really must honour. Then, there is your financial advisor, keen for you to review the latest set of accounts and make important decisions accordingly. Nor must we forget your chief of justice. This very important figure doesn't just want to make sure that all is fair and proper in your own world. The head of law and order also feels determined to engage you in discussion about the appropriate rewards and punishments for all kinds of people in the wider community.

Many of your opinions, and sometimes your prejudices, too, are formed during these daily consultations with this aspect of your own personality. It's not for me to judge whether *you* judge too often or too harshly. Nevertheless, I cannot help but notice that it is not just others that you are sometimes very quick to declare 'guilty'. You also, it seems, have a tendency to dwell unnecessarily long on your own guilt.

Watch for your blind spot

It would appear that, somewhere within your carefully governed, independent state, you have a 'Department of Self-Denigration', run by a very dour and disapproving character who follows you around, reciting a long list of your faults, failings, mistakes and errors.

If you're in a good state of mind, you can limit consultation with this part of yourself to an essential minimum. You *do* need to know about your mistakes ... but only in order to learn from them, and not so that you can dwell on them. Your faults definitely ought to be acknowledged. But acknowledgement involves a polite, passing acceptance – not a deep, detailed and overly intimate analysis.

Usually, when I can't get through to you because

you are feeling too guilty about something, I find you are also listening to another part of your complex, multi-faceted persona.

Amongst your many resources, you undoubtedly have an inner warrior, but far too often you pay more attention to your inner worrier. It stands before you with a serious expression on its face, reciting a long litany of problems, troubles and potential sources of deep disaster. If you are not careful, you can end up spending day after day consulting that scary list and finding scant time to notice anything else.

What if you refuse to consider it? Well, we have to ask whether this is because you have decided to make time for a talk, instead, with your inner force of trust and faith, or whether you have simply opted to listen to your psychological spin doctor, who is rapidly helping you create a series of reasons to ignore the facts that need to be faced.

In theory, as your guardian angel, I am more powerful than the rest of these inner forces put together. In practice, I am only as useful to you as you allow me to be. No matter how good my advice is, it won't help you if you don't stop and listen to it. And if, in your life, you are in 'denial' about anything or anyone, you are never going to listen to me or anyone else. You are going to be rather like a motorist, driving a fast car with a dangerous blind spot in one of the mirrors.

That blind spot will be – or rather, *is* – your 'addiction'. It is not possible for a human being to exist on planet earth without an addiction of some kind to some thing. Addictions are very raw and real things. They feel fine while they are being fulfilled ... but when they're not, people become very edgy and unreasonable. In one way, it doesn't much matter whether the addiction is to coffee or cocaine, heroin or heroism. The world may approve of you for having one kind of habit and disown you for having another, but wherever you've got a powerful need that keeps on overwhelming you, you've got a mighty soft and vulnerable spot inside yourself that's worryingly open to the wrong kind of persuasion.

Whatever you are hooked on, you are hooked on it more strongly than you realise. Whenever or wherever you are deceiving yourself about something, I am at a disadvantage and so, ultimately, are you.

Why should that matter? Well, remember, you and I need to work as a team. When we both pull together in the same way, we can make amazing things happen. I can identify the opportunities and you can go and seize them. Or, if you prefer, you can cook up the plan and I can help gather up the possibilities that will help to make it a reality. Either way, it all works a lot more smoothly when we are reading from the same page.

I don't mean to imply that I can't help you unless you somehow overcome all fear, ignorance, prejudice, negativity and self-delusion. You can aim for this if you like, but you had better make sure that in the process you don't end up kidding yourself twice as badly as you were doing in the first place.

Perfection is not the aim of the game. Understanding, though, *is*. If you're wise to yourself, aware of your foibles and conscious of your tendencies, at least you can compensate for them to some extent. At least you can laugh at yourself – and at life in general. At least you can enter into constructive, intelligent dialogue about difficult issues rather than hiding or running away from them. And at least, even if you still have your bad habit, you can put it where you and the rest of the world can see it. You can stop it from covering itself in a cloak of conspiratorial secrecy, and then obliging you to spend vast tracts of your time hiding the truth from yourself . . . or others.

As your guardian angel, I know a great deal about where your deepest personal cravings are, what sparks them off, what you feel you need to feed them, and how far you are prepared to go to satisfy them. *'This is what I need'* your addiction makes you say, *'and I need it right now!'*

Fortunately, as I have said many times before, I don't judge. I do, though, sometimes wish for things and one of the things I'd most definitely ask *my* guardian angel for, if I had one, is the ability to help you learn a little more about yourself.

You do certain things without admitting to yourself that you are really doing them. You tell yourself, for example, that you feel fine about situations that actually make you feel upset or angry. Sometimes you help yourself to comfort foods and then try to pretend that you didn't eat them. Or you act obstructively to people that you dislike whilst smiling and telling yourself that you are just being normal and reasonable. All these things are perfectly normal and natural. No human being on earth doesn't do this from time to time. But most just insist that it isn't happening, while some at least make an effort to 'catch themselves' in the act. If only you could strive for a little more perspective on your own unconscious behaviour patterns, you'd make an enormous difference to your ability to enjoy life, to be successful, to avoid repetitious and negative situations . . . and to work well with your guardian angel!

I can show you what you need to see about yourself

You could, of course, try asking me to help show you what you need to see about yourself. However, one of the main reasons why you are usually reluctant to seek such enlightenment is that you fear it would fill your life with an uncomfortable amount of doubt.

CHAPTER 5

The glory of doubt

Doubt, although it can be destructive, is rarely anywhere near as dark or dangerous a force as certainty. Lots of people, especially in the New Age communities around the world today, describe themselves as 'Light Workers'. They say that they are working exclusively with the forces of positivity and harmony, hope and spirituality.

I never like to be rude, but often I secretly want to tap them on the shoulder and ask them, 'How can you be so sure?' – because light, in my experience, always casts a shadow. And very few lights, in this world, cast shadows as dark as the bright beams of certainty.

Although I never judge, I do sometimes wonder whether certainty isn't one of the most evil forces on the face of the planet? It often seems as if it is where most negativity, ignorance, cruelty and insensitivity come from.

Strangely, it seems to be at its most potentially poisonous when it is married to spirituality. Nothing does more damage than an absolute insistence that right is on your side, that your god is talking to you, that you know exactly how everything needs to be ... and you can see where everyone else is going wrong. Most of the bloodiest, most barbaric events that have ever taken place in history have been based on an unshakeably deep faith in some 'higher purpose'. The worst conflicts of all invariably stem from two groups with different beliefs each setting out to prove that ultimately, the Supreme Being is on their side!

If you are sure about what's right and what's wrong, in your life or in anyone else's, then, no matter how full of golden virtue your ideas may sound, you are actually a perpetrator of prejudice and a deepener of darkness. Doubt, then, is surely better than getting on a high horse and riding roughshod over reason, trampling tolerance, crushing compassion ... and undermining understanding. And this may well be why often, when I need to get a message to you, I tend to do it by inviting you to doubt something that you have previously accepted without question.

I send you as many signs and signals as I possibly can. I try every trick in the book to let you know what I feel you need to know. You are, however, a very busy person with a lot on your plate. No matter how open-minded and flexible you try to be, you tend to have a lot of fixed ideas. You can't help this, but whenever a new development occurs or a fresh piece of information makes its way into your world, your immediate instinct is to use it somehow to reinforce the world-view you already hold. It doesn't much matter whether this is a pessimistic view or an optimistic one. It's the view you've got and there is a good reason why you usually feel inclined to hang on to it.

Life is full of mysteries. You don't know where you were before you came here, you don't know where you will be going when you leave. Nor do you know exactly what you are supposed to be doing with your time on this planet. You don't know and nor do your fellow humans, however deep their convictions might be. That's a somewhat unsettling proposition. It's enough to make even the most confident person feel a little insecure. A very natural reaction to this involves fabricating a life raft of assumption and opinion in order to afford some protection from the endless waves of uncertainty that could otherwise threaten to engulf you at any moment.

To put it another way, everything in this world is changing all the time. Nothing ever stays the same for very long. Some things give an appearance of permanence – buildings, for example, or trees, or nice, solid pieces of furniture. Nevertheless, all these things are slowly decaying and disintegrating. They may be capable of lasting a good few thousand years, but they are still little more than fleeting blips on the radar screen of time.

To be a human being is to be an infinite, eternal spirit in a finite, temporary form. It is to be perpetually seeking the one thing that can never be attained. Permanence. You feel you need this yet, somewhere deep inside, you know that you cannot have it. Thus, you focus on the things that seem most stable and draw what comfort you can from them.

So far, so good. Unfortunately, it doesn't stop there. At the essence of your being is a contradiction in terms. You are finite, yet infinite. Temporary, yet eternal. No wonder, then, that you are contradictory in other ways and that your life is full of contradictory impulses. You crave security and stability, yet you also yearn and burn for change.

Invariably, when you ask me for help, that help has something to do with your need for 'change management'. Either you want to keep a change at bay or you want to bring it on and welcome it into your life as quickly and fully as possible. All the time, from the throne room of your psyche, you are scanning the horizon. Through your ears and your eyes, through your intuition and your intellect, through your heart and your head, you are attempting to interpret the signs and signals that you see all around you. What is each new development trying to tell you? What does it really mean? Most often, you incorporate the information into your existing perspective in a way that helps you hang on to what you think you already know.

Generally, however, my job as your guardian angel involves introducing you to the new and the different. Although I often send you signs, you don't usually see them. But every so often you will get a feeling so powerful and overwhelming that you simply cannot ignore it or subject it to any form of scrutiny.

Then, you are fairly safe to assume that we are making contact.

The paradox is that if you want to increase the frequency of successful communication between my realm and yours, you should also be suspicious of whatever you think is absolute and indisputable. Therefore, the other time when you can be pretty sure that you are reading me right is when you really aren't sure whether you are being told or shown anything at all!

So, signs and signals can often be confusing no matter how much clarity they may appear to bring at first. Happily, there is another way in which you can improve your ability to hear what I am saying to you. You can simply ask me to help you to be more receptive. That is a very simple but powerful request. If you ever make it, you will be rewarded with the most wonderful results.

Undoing what has been done

For the record – and because it needs to be said – I have absolutely *no* problem with you changing your mind. About *anything*. You don't have to worry about wasting my time. I am, after all, an eternal being. I have all the time in the world. But you don't.

Sometimes, the only way you can learn valuable lessons is by making mistakes. By trying something out, discovering that you don't like it or that it doesn't work for you ... and then realising what you actually do want. I completely understand this. I don't hold it against you if you ask me for something one month, and then ask me to help you lose it from your life the next, or if you change your mind in the course of a single day. It doesn't bother me. It may bother you, though.

Variety is the spice of life. But changes that happen faster than a stroboscopic light are often just the source of a big headache. And while I'm happy to endeavour to provide you with as much change as you ever request, as often as you want, it's worth remembering this: a big change might make your wishes appear totally illogical or inconsistent in the eyes of others. If you want to avoid such criticism, just don't tell the people who are likely to judge you. Tell me instead, and remember ... I don't judge.

Nor do I ever deliberately misunderstand you. I don't 'accidentally on purpose' mishear you. I don't twist your words. I don't seize the opportunity to teach you a lesson. I play with a straight bat. I do just as I am asked. I don't read 'a little extra' into any request that is ever made of me.

It is true, though, that if you are very, very specific about what you are asking me for, I am likely to be very, very specific about what I try to supply. This means that I won't offer you a substitute, because you are not asking me to look for one. It also means that, even if I can see the one thing that you *ought* to have added to your order but have clearly forgotten, I won't automatically put it into the package.

I do just as I am asked

If you want *that* kind of conscientious, caring service, you are most welcome to it. All you have to do is ask me for my recommendation. Invite me in on your decision-making process. Allow me to make suggestions. Express (and display) some willingness to be guided.

CHAPTER 6

Positive thoughts, positive gestures

I have already explained that rites and rituals don't especially impress the residents of the ethereal realm. We guardian angels don't have a 'preferred way' of being approached. It's all the same to us if you speak in deeply religious terms, or chat to us like long-lost buddies. What matters to us is that your way works for you.

However, it is obvious to us that it is *not* all the same to humans. Oh, the squabbling that goes on, the competitive one-upmanship, the holier-than-thouness, the pompous, self-righteous justification, and the heartlessly dismissive cruelty that occurs in the name of deep belief. Oh, too, the embarrassed apologies we receive, the heartfelt exhortations we hear, the humble requests for forgiveness we get.

Still, we do not judge. Can you imagine what would happen if we did? There would be no need for courts of law in your world. The all-seeing, all-knowing team of angels could dispense it from theirs. So, if someone ever committed a crime, there would be no need to track them down and put them on trial. We'd just take away their breath and let them die on the spot.

And if someone did a great deed? The department of angelic justice would ensure that nothing 'bad' ever happened to that person ever again in their life. Does it work that way? Yes and no.

There is, of course, such a thing as 'natural justice'. People who seriously step beyond the boundaries of what's okay find the cosmos has a way of catching up with them, regardless of whether or not the law ever does. But it happens automatically ... through the individual's *own* inner process. Someone who knows they have been 'bad' carries that knowledge with them in their heart, everywhere they go, through everything they do. Eventually, one way or another, it gets to them. Likewise, those who have been consciously 'good' get their reward, too ... in the way that they then, somehow, attract other positive forces into their lives.

Guardian angels really don't have much to do with any of this. I know this is going to sound shocking, but if you turn, tomorrow, into a mass murderer and commit some of the most abominable crimes in the history of humanity, I'm absolutely not going to judge you, punish you, or desert you.

From a conventional moralistic point of view, it gets worse. If you sincerely ask me for help in committing those crimes ... I'm going to aid and abet you every bit as far as I am able. If you ask me to help hide and shelter you once you are on the run, I will gladly offer you my cosmic assistance. If you want my support in creating an alibi to 'prove' that you were somewhere else at the time, I'll willingly engineer as many convenient coincidences as I can.

Really? Yes. Sorry. Really. It's true.

There are, however, certain things I can't and won't do. We touched on some of these in earlier.

I can't do anything that puts me in immediate direct conflict with another person's guardian angel. So, if you'll excuse the graphic imagery, I can get you the knife, and find you the alleyway to hide in ... but if you want my help in leaping out of the darkness to attack, you can't have it. And that, by the way, is regardless of whether your victim is entirely innocent or in fact the evil perpetrator of some far more dreadful deed, who, in your opinion, entirely 'deserves' payback.

Guardian angels don't judge ... and they don't do conflict. Whenever human beings are deeply engaged and engrossed in either of these activities, guardian angels take time out. Sadly, a lot of people spend a lot of the time either judging or fighting each other. Which means they spend a lot of time being ostracised from their angels.

Then again, it's not all bad news. It means that if you ever want more interaction with your angel, all you have to do is stop judging ... and stop fighting!

Affirmation and expectation

Another way you can get me to do more for you is by adjusting your frame of mind – and by paying attention to your own level of expectation.

On the one hand, it really helps me if you have a positive attitude. If you don't really think that I can do anything for you, the likelihood is that, even if I engineer some extraordinary coincidence, you'll fail to notice. You'll be suspicious of even the most amazing opportunity. And even if you accept it, you may not do so wholeheartedly. You won't get as much from it as you could.

You are much more likely to get success if you start acting and talking as if you *expect* it. Your confidence will prove contagious. Others will believe in your ability to do what you say you are intending to do. Or to get what you feel sure you can get. Oddly, they will start looking out for ways to help you. (Or, at least, they will if they are positive, too. We'll deal with negative companions in a while.)

you need to be open-minded

You, in turn, will feel better just for knowing that soon you're going to feel better. If you want, you can even start saying things such as, 'I *am* a person with a nice home / happy family / good job / great talent / loving partner / real sense of purpose / plentiful income, etc.' Even if these things have yet to happen, you can show your trust in the notion that they will happen and, as you do, you can make it more likely that they will become realities.

There may be many an occasion when the best I can do for you is to steer you in the direction of a 'good chance'. It's actually quite rare that I can fulfil a cosmic order 'on a silver platter', exactly as requested.

The very fact that I can't always provide *exactly* what you think you want is the reason why you need to build a little flexibility into your expectation. This means, at the same time as definitely expecting something good – *and* having a fairly clear idea in your mind about what that might be – you also need to be vague and open-minded enough to suddenly see that I'm offering you a 'substitution' for your order. When you first start to think about it, this seems like a tricky balancing act – but once you get the hang of it, it's like riding a bike. You'll never forget how to do it.

A substitution might not just be of 'equivalent value' – it could even be worth a whole lot more to you. There's nothing wrong with wanting a very specific thing, such as the widget for the car I mentioned earlier. But if you ask me, at the same time, for a 'nearest equivalent', you make it much easier for me to fulfil more of your requests, more often.

One way that some people make this work with their angels is by using the phrase 'putting out'. They say, 'I'm putting out to the universe / to the cosmos / to my angel for an assistant with this project or for a solution to this plan.' Or for a better idea. Or for a way to get this particular problem solved. Or for a stroke of inspiration. Or for a lucky break. Or for something that makes it all easier for everyone involved. They deliberately leave the details undefined. That way, we angels get a little creative input!

What about other people's negative input?

If you're asking me to help you with something, should you tell anyone else? That depends on how much you trust them ... and me. There certainly is no *need* to share the details of our very personal, intimate relationship. Yet nor do you have to feel ashamed of it. Telling someone else about what you are trying to accomplish or attain could help them to be more aware of what you need and thus be more able to help you. However, it may also enable them to stand in the way of your progress. If that person disapproves – or feels insecure about your objective – or just has a downright malevolent agenda, they may try to do what they can to prevent me from helping you.

There's an easy way to stop that from happening. Ask me to alert you to anyone you need to be especially discreet or careful with. This is one of the easiest and most successful ways we can interact. Ask me to show you some sort of sign or signal whenever you're talking to someone who isn't as 'on your side' as you might imagine. Request a subtle little hint, like a funny feeling on the back of your neck, or a strange little tremble in your tummy.

Likewise, try asking me to help you recognise the

folk who are most likely to support or help you. Some people have a funny way of saying 'all the right things' even when they are doing all the wrong ones. Others can come across as very gruff and negative, even when they are actually of kind heart and good spirit with plenty to offer you in your current situation.

The way to distinguish the difference is to 'filter your hearing'. Listen to what's being said to you, of course. Notice the tone of voice, naturally. But then go one step beyond and try to 'read' what's in their heart.

read what's in their heart

If in doubt about whether you are picking up the clues clearly, ask me. Processes that involve me working directly with an inner faculty in you invariably work better and faster than those that require external intervention. Make one sincere request to me for better insight into the true motives of others, and I'll keep on supplying you with this until or unless you ever say 'stop' – or you become overwhelmed by a perverse and powerful desire to be deceived.

The roles people play

Why might you ever experience such an urge? Because you live in a world of icons and images. Shakespeare wasn't wrong when he said that 'All the world's a stage'. You may not think of yourself as a thespian, but a lot of the time you cannot help but play out roles. Parent. Child. Lover. Spouse. Employee. Boss. Customer. Supplier. Expert. Amateur. Enthusiast. Cynic. Creator. Consumer. Driver. Passenger. Victim. Antagonist. Defender. Attacker. Teacher. Student. Leader. Follower. At various times in your life so far, you have been obliged through force or circumstance (or choice) or both, to play out almost all of these parts ... and at moments of great intensity, you have thrown yourself deeply into one role or another.

at moments of great intensity, you throw yourself into a role

When you're totally engrossed in 'acting out' a part, all you want are the support mechanisms necessary to help you play that part properly. If you're trying to be a student, you want someone who will teach you. If you yearn to be a lover, you must have someone to love. And if you're being a parent ... well, please bear with me here for what I have to say is a little difficult to explain – but extremely important to grasp.

No two children in this world are the same. No two parents, either. But there are some situations that are pretty much 'universal'. When you're looking after a very small child, on your own, it doesn't matter where you are, in which part of the world. Nor does it really matter how much money you've got, what kind of an education you have had, who 'else' you are when you are not in that situation – you are currently fulfilling just one function. There will come a moment when, no matter how much you love the child or how deep your relationship with them has been, your concentration wanders and your patience, temporarily, goes missing. At that point, you'll say and do pretty much whatever any stressed parent says or does in a situation of this nature. And the child will react in much the same way as any other child might, regardless of intellect or individuality. You will both, at that moment, be acting like stereotypes, cartoon caricatures, clichés.

There are points, too, in any adult-to-adult relationship when much the same thing will be going on. If this relationship also involves a child, it is almost impossible for everyone involved not to fall into two-dimensional, cardboard cutout, stereotypical roles from time to time.

Parent–child. Child–parent. Parent–parent. Unconsciously, people tend to behave (especially when under stress) as they have seen others behave. As their parents behaved when they were being brought up. As they have seen other parents – or other children – behaving. What's wrong with this? Nothing. But it can skew their judgement sometimes.

If you, for example, are playing the role of parent, you may, perhaps, feel a very strong need for someone to play co-parent with you. You may have such a person in your life already – but you're finding that the relationship is less than ideal. Or you may simply yearn to have an ideal person come into your world and fulfil that very particular need for you.

Whether you ask me for help, or whether you just start 'looking', you will find yourself almost unable to resist eyeing up potential candidates for the role that so much needs playing in your life, to complement the role you are playing. If someone comes along who looks as if they will work well in that one area where there is an outstanding, urgent, deep, intense, aching, echoing need ... you will forgive them a thousand faults and overlook a million other points of potential incompatibility. You will deceive yourself almost willingly and gladly, just for the sake of getting what you're after in the area where you feel the strongest sense of 'want'. Worse, the other person may do the same.

They may not be interested in who you are at all. Their attraction could be almost entirely towards 'who you are currently busy being'. Like movie fans who can't tell the difference between the person they have seen on screen and the actor on their day off. They may provide you with much interest and interaction – even admiration or adulation – but only because of your apparent match with their greatest inner desire. They may be deceiving themselves about you. You may be deceiving yourself about them. The two of you may get caught up in an extremely convenient process of mutually compatible delusion ... and it may all be fine until eventually, for one or the other of you, the bubble bursts.

they may provide you with admiration

There *are* times when all humans deliberately deceive themselves. Or rather, when they unwittingly allow their current needs and desires to blind them to the point where they cannot see a bigger picture. As a guardian angel, I understand this very well. I just sometimes wish you understood it a little better. Especially when you were busily asking me for things that were not what you really wanted ... but just things that would ultimately help to maintain a state of self-deception.

CHAPTER 7

Perspectives and priorities

One thing that I can be particularly helpful with is in 'establishing priorities'. That, I have noticed, is often very difficult for humans to do. Some things become of seemingly crucial importance, even though it's debatable how much they actually matter. Other things, that might really warrant far more attention and energy, seem to get pushed to the back of the queue.

I don't distinguish between big and small or vital and irrelevant in quite the way that you do. But I do know how important all this is to you ... and how uncomfortable your life can become when your 'relative values' are not in an order that really makes sense to you. Indeed, this reminds me of yet another story.

* * * * * * * * * * * * * * * * * * * *

Once upon a time, there were two travellers. They were just wandering together from village to village and town to town, looking for new adventures in the game of life. They had made many successful visits and had some most enjoyable journeys. Eventually, they came to a land that neither had heard of before.

As they arrived at the outskirts of a town, they stopped a local person and asked where they might get food, water and shelter for a night or two.

'Right here, folks,' came the reply. 'I've got a wonderful meal you can eat, some magnificent wine you can drink, and the comfiest beds you have ever slept in.'

Something about this reply made them slightly nervous. 'We haven't got much money, you know . . .' said one of the travellers. 'Our board and lodging needs to be modest and humble.'

'Well,' said the local resident, 'you're going to like it in this part of the world. We have a system of payment and purchase . . . but it's very different to the kind that you're used to. We don't have notes or coins of different denominations. Round here, everything you want, no matter what it is, costs exactly the same amount. Just one penny.'

The travellers could hardly believe their ears. 'A penny? No matter what it is?'

'Sure,' said their new-found friend.

'Oh, come on,' said one of the travellers. 'What if I want a loaf of bread?'

'Then, you'll have to pay a penny,' answered the local.

'And if I want to buy a diamond ring?'

'That's a penny, too!'

Somewhat taken aback by this information, the two travellers looked at each other in amazement. Realising that they at least ought to have something to eat and drink while they let this all sink in, they asked for a meal. Shortly afterwards, a plateful of truly attractive, appetising fare arrived in front of them. They were very hungry, but before they picked up their forks and began to eat, they felt they had better check once more.

'How much do we owe you for this?' they asked.

'Just a penny each,' came the reply.

The food tasted even better than it looked. For a further penny apiece, they purchased something delightful to drink, and proceeded to discuss the situation.

'Well,' said the first traveller, 'I reckon we've come to a wonderful place. It's hard to imagine, isn't it, how we're ever going to want to leave this land?'

'Is that really what you think?' asked the second traveller. 'Personally, I don't like it one bit. In fact, I want to get as far away from here as possible, as fast as I can. And I'll be very happy if I never come back.'

'What's your problem?' said the first. 'All that travelling must have addled your brains and made you crazy.'

'On the contrary,' said the second. 'I can hardly believe, after all the places you have been and the things you have seen, that you don't share my instinctive dislike for this extremely peculiar and potentially dangerous situation.'

They argued for some while, each finding themselves entirely unable to comprehend the other's point of view. Eventually, they reached a decision.

* *

Unfortunately, before I tell you what this was, I have to interrupt the narrative flow for a moment. So far, in this tale, I have delibertately dodged giving either of our central characters an identifying feature. I have so carefully avoided description and definition that I have not even told you whether the travellers are male or female.

As a guardian angel, I dwell in a realm where such distinctions are irrelevant. But I fully appreciate how much they matter to you and that is even more reason why I am reluctant to emphasise them. The moment I tell you what the nationality of these travellers is, or give you details about their cultural background, I'm going to limit your imagination a little. As things stand, you've already instinctively conjured a picture in your mind about these characters. If I asked you, right now, to paint a picture of the two travellers, you could. You could also describe the local resident they have just purchased the meal from, the surrounding scenery, the weather and the temperature.

You may not be consciously focusing on any of this detail, but inwardly you've got it all to draw on if you need it. Human creativity is an awe-inspiring thing. It 'fills in the gaps' for people all over the planet, all the time. It assumes and it expects. It guesses and it figures. It takes what it knows and adds, to this, what it would 'like to know'.

If you are relatively sympathetic to the tale I have told you so far, you have probably made those two travellers friendly folk. You have dressed them in the clothes you would like them to be wearing if you were ever to encounter them in reality. You have given them facial expressions that you can sympathise with and accents that sit easily on your ear. You have made them your people. In the few short moments since I first mentioned their existence, you have adopted them, embraced them and adjusted them to suit your own particular preferences.

That, of course, is all absolutely fine – as is the fact that now you would rather like me to stop waffling on and get back to the story so you can find out what happens next. Before I do, though, let's just look a little longer at this tendency that you have to create an unspoken, private 'back story' full of additional details. It is a trait that can, sometimes, add a level of complication and confusion into your dealings with me.

As you know by now, I always respond willingly and happily to any attainable request that you make of me. My only requirement is that you are sincere. If you add a lot of detail when you make your request, I take it on board and attempt to be as exact as I can. If you are vague – and especially if you allow me some creative freedom of my own – I can work with the spirit of what you are asking for and supply something that I can be confident will work for you.

You *ought* to be able to trust me, because I can hardly be your guardian angel without knowing you pretty well. I won't overstep my instructions unless you specifically tell me that this is what you want me to do. And here is where, if we are not careful, we can wind up wandering into a troublesome grey area.

I can read your mind. I really can. I can hear your every thought, I can see your every fantasy, I can sense your every need ... and I can detect your every desire. There's no need for you to be embarrassed about whatever you are thinking, or hoping, or wishing for.

I have a sneaking suspicion I may have mentioned this before, but just for the record, I don't judge!

Yet I can read what's in the back of your mind just as easily as I read what's at the forefront of your consciousness. Often, indeed, I know more than you yourself know about the ideas you are secretly entertaining and the options you are privately contemplating. I have to know this. It's my job. I'd be a pretty useless guardian angel if I didn't. Even now, as we speak, I'm conscious of the fact that a part of you is getting a little physically uncomfortable. It may be a good idea to get up and stretch your legs for a while before you read too much more ... You may also need to attend to the needs of your digestive system while you're taking your break. Feel free to put the book down for a few moments if you want to. But do come back soon. The next point that I want to make is a pretty crucial one.

Hidden agendas

Okay? Right. Here's my point. What am I supposed to do if you say to me, one day, 'Please help me get a holiday?' Suppose, for example, you ask me very specifically for a ten-day break by the side of a particular beach, in a hotel of a certain kind and quality, in the company of a very specific individual?' Suppose I manage to come through with a complete, exact fulfilment of this very precise cosmic order? What might stop you from feeling entirely ecstatic about this?

Well, suppose you have consciously concentrated on your request for all of the above ... but all the time, under the surface, you have been hoping that the holiday will also involve an interesting expedition, an enjoyable social encounter with a new friend, a chance to play some sport you have never tried before, or a visit to a type of restaurant that you much enjoy.

Do you want me to *notice* all the detail that you haven't specifically gone into? Or shall I ignore it on the grounds that you haven't actually mentioned it? You might consider it rude and invasive of me to go probing into areas of your mind that you have not officially invited me into. Or you might just feel disappointed by the fact that I have stuck rigidly to the letter, rather than the full underlying intention, of your request.

Ideally, I would love you to be vague and say 'surprise me, nicely'. But I fully understand why you may want to put in a clear, exact request ... and if you do, I would very much like you to clarify how far you want me to go with it.

I expect by now you may be thinking, 'What kind of madness is this? Why on earth [or anywhere else] would I *not* want my angel to hear my quiet unspoken wish every bit as clearly as my formally expressed and articulated desire?'

But suppose you've got a half wish that is a little counterproductive? Suppose you half want to go on holiday with your partner – and you half hope your partner will stay at home and let you go alone? Suppose you half wish you might fall in love with a stranger on this break – and you half wish that you will have an entirely simple, easy time to yourself? Suppose you half wish you weren't even going on that holiday because you also half wish that you could just stay at home and find a way to enjoy the very thing that you think you need a break from?

All I'm saying here is that your dreams, desires, wishes and aspirations are not always as straightforward as you imagine. It's really not such a great idea to make requests to your guardian angel if you're not entirely sure, deep within, about which requests you want to be fulfilled and which you want to be ignored. It makes it extremely difficult for me to help you, not least because I'm in danger of disappointing you in two different ways.

I can let you down by *not* paying attention to those

'unofficial extras' on your list. Or I can let you down by paying them *too much* attention and giving you something you wish you hadn't had!

Clarity on your part is what I'm asking for. That, and a carefully thought through sense of perspective and priority. All of which, finally, brings us back to our story.

For the sake of an easier narrative, I want to give our two travellers a sex. But if I make them a man and a woman I'm slightly concerned that this will imply a romantic relationship. Traditionally, this story is told with two males in the central roles. There is, though, really no reason why they can't both be female, so I shall make them women and have done with it – women travellers, seeking wisdom and depth of experience in a long journey around the world.

And, as we need a little language to distinguish one from the other, I'll call one the older woman and the other the younger woman. (Not, by the way, that I am automatically attempting to imply that age and wisdom are in any way related. Words can sometimes make communication very difficult in your realm.)

Anyway, here we are, back with our two travellers – debating the pros and cons of a world where everything costs a penny.

* * * * * * * * * * * * * * * * * * * *

'Really,' said the older woman, 'this place strikes me as trouble. I want to get as far away from it as I can, as fast as I can.'

'What's your problem?' said the younger one. 'It's heaven on earth, that's what it is. We can have the time of our lives here.'

'The time of your life, maybe. But not mine. You stay here if you want to. I'm going on, alone if I must.'

'I can't believe you're saying that. I absolutely have to stick around long enough to find out how things work here,' replied the younger traveller.

The next morning, they woke up to discover that they were still of an entirely different opinion and they agreed to go their separate ways. The younger woman was keen to go off exploring. She had a fine time in the many bazaars and market places. Her pennies went a very long way – and she was much enjoying all her new possessions. After a while, she decided she was hungry again and, as everything cost a penny no matter what it was, where you found it, or who you bought it from, she headed off for the finest restaurant in town.

On her way she passed a noisy altercation involving several people standing in the street. They were arguing very heatedly and she couldn't help but overhear their discussion. A goat, it transpired, was lying dead in the road.

Its owner was furious. 'My animal was precious to me and now it has been suddenly and cruelly slain by a stone. As we all know, goats are sacred animals in this city and the punishment for anyone who is ever responsible for killing one of these noble beasts is death. We must know who threw the stone.'

A passer-by tried to explain that he had witnessed the entire event. 'Nobody threw the stone. It fell from the balcony above.'

'In that case, we must execute the owner of the property.'

Another voice then entered the fray. 'I am the owner ... but it is not my fault. Only recently, I paid a master builder to have that balcony made safe.'

The mob became angry and vociferous, attracting the attention of a royal delegation.

The King was out walking through the city, accompanied by several senior officers and advisors. His Majesty was keen to know what had happened and to establish who was responsible. He ordered his troops to go and fetch the builder.

As he was being marched towards the gathering, the builder started protesting at the top of his voice. 'I built that balcony, it is true ... but it was not my fault that it collapsed. I gave the job of cementing those stones to my colleague.'

The colleague in turn was called for . . . and he told his story. 'I mixed the cement, it is true. But it is not my fault that I could not do this with as much water as I should really have used.'

'Why?' asked the King. 'Whose fault was it, then?'

'Well,' said the man, 'at the crucial moment when I needed to fetch more water, my way was barred by one of your soldiers. This one here, as a matter of fact . . .' he said, pointing to one of the King's senior officers.

'Is this true?' inquired the King.

'Your Majesty, it may have been,' said the soldier. 'For I remember stopping this gentleman a few days ago when you were leading a royal parade through the city. You distinctly told me at the time to make sure that nobody went about their normal business until your carriage had passed along the street.'

'Oh,' said the King. 'Then it must be my fault!'

A hush fell across the entire crowd, followed by much whispering and muttering. It was generally agreed that the King himself could not be executed . . . but that somebody must be made to pay for the death of the goat. The question was – who?

At this point, a few people in the crowd began to notice our traveller, and before long, several hundred people were all marching towards her, pointing their fingers accusingly.

'It is your fault,' they said.

'How can it be?' she asked.

'Because we say so,' came the reply. 'And because there must be a punishment.'

From the cold discomfort of her condemned cell, our traveller had plenty of time to reflect on the unfortunate implication of passing through a land where everything costs the same – and nobody can put a real value on any transaction. Even when it comes to dispensing justice, nothing is worth more, or less, than anything else ... with the possible exception of the King's life!

* * * * * * * * * * * * * * * * * * * *

Now, I hope, you can see why I wanted to tell you this story. When it comes to cosmic ordering, priorities matter a lot. As do clear comparative values. Without these, there's a real and horrible risk that in the quest to ask the universe for everything, you could end up worse off than you would have been had you asked it for nothing.

But now, I really need to tell you the end of the story, because it too has a moral that could prove extremely enlightening.

* * * * * * * * * * * * * * * * * * * *

It was a few days later – and hundreds of people had gathered in the square to attend the show trial and the summary execution of our unfortunate heroine.

She was doing her level best to be stoical and brave with what might be her last moments on the planet. Deep within though, she was extremely unhappy and regretful. Now she understood why her friend was so keen to get away from this place and she wished she could at least see her companion long enough to acknowledge this – and to say farewell. But at the same time she hoped her friend was by now far away. She'd hate her to meet the same fate.

Her friend, though, was actually very near. The older woman had not travelled far when news reached her of her young companion's unfortunate situation. At that very moment, she was edging her way towards the town square, accompanied by an entire herd of prize goats. Each of these she had purchased that very morning. Alhough she didn't have much money, she had been able to afford them because they were, of course, just one penny apiece.

As the soldiers led her friend towards the gallows, the King began to make a speech about the need to punish wrongdoing. Next to him stood all the people who had been involved in the fracas on that fateful morning ... including the owner of the deceased goat.

The older woman forced her way through the crowd till she reached the front . . . and then she shouted out, loudly: 'Stop! I have a proposition to put to His Majesty.'

An officer made as if to silence her, but the King held him back. 'Let her speak,' he said.

'Your Majesty,' said the woman, 'may I ask you a question? In your great and wonderful kingdom, everything costs a penny, does it not?'

The King nodded in agreement. The woman went on:

'So, Your Majesty, what is the value of this apple I am holding in my hand?'

The King replied, 'A penny, of course.'

'And of the crown you are wearing on your head?'

'A penny.'

'What then, Your Majesty, is the value of the goat that died when the stone from the balcony fell on its head?'

'A penny.'

'Thank you, Your Majesty. My questions are nearly at an end. Firstly though, I must ask you about the human life you are about to end for the sake of a stone falling on a goat. Is this woman's entire existence worth merely a single penny, too?'

The King looked pensive for a moment and then he stiffened with inner resolve. 'If you choose to see it that way, I cannot argue. But now, you have delayed the execution long enough. May we continue?'

The woman said, 'I had a feeling you might say that, so I would like to put a question to the owner of the goat.' She turned to look at the goat owner. 'Would you accept this entire herd of goats in compensation for the death of your single goat?'

The man looked at the King. The King looked at the man. They both looked at the woman. The King spoke, gravely, 'In our land, madam, absolutely no system of compensation exists, other than the trading mechanism by which everything costs a penny.'

The woman replied quickly, 'Well then, would the goat owner be willing to buy this entire herd from me for a single penny?'

The goat owner looked very keen to accept. Not waiting for a reply, the woman went on, 'And Your Majesty, as you have already agreed that this young woman's life is worth only a penny, too . . . may I pay you that penny in return for her keeping her life?'

The King looked nonplussed for a moment. The young woman's heart began to race. All the hope that had drained so sadly away from her over the past few days began to flood back in. She looked at her friend – and at the King. 'I can't believe it!' she thought. 'I really am going to be saved.'

The King looked directly at her. He then turned to her friend at the front of the crowd and said, 'Sorry. Too late.

The sentence has been passed. There must now be an execution.'

The young woman hung her head in despair and disbelief. For one wonderful moment, her faith had been revived and restored. Now it was flowing away from her faster than the water in a broken dam.

The older woman was mortified, too. Her big plan had been to step forward, hand the goats over to the goat's owner, demand a penny from him, press that same penny into the King's hand, grab her friend by the arm and whisk her out of there. Now she was utterly defeated.

For a split second, all she could think about was the horrible reality of her situation. Her powerlessness. Her hopelessness. Her inability to do anything other than stand and watch while these desperately unjust strangers killed her friend before her very eyes.

Inwardly, automatically, almost entirely unconsciously she reached out to her guardian angel. 'If there's anything I can do – or you can do – to change this, please let it happen right now!'

Suddenly, she found herself acting and speaking without any idea of what she was doing or saying. It was as if she was being taken over by some powerful, inexplicable force. As if somehow, she was watching herself and listening to herself from a distance while something else took over control of her body.

She heard herself shouting back to the King, in a confident, authoritative tone: 'In that case, don't hang her ... hang me!'

The King looked at her. Was she serious? The woman pushed her way forward. 'Do you not know? Today is the day of a great cosmic alignment. Have you not heard? The portal to heaven is wide open. In accordance with ancient prophecy, the first person to die today will spend the rest of eternity in a state of divine protection and forgiveness, regardless *of what life he or she has lived.'*

The King and his courtiers looked at each other. Was she mad? Or was this true? The woman, meanwhile, continued to push her way through the crowd shouting, 'Hang me instead. And please, be quick about it.'

The officials on the stage may not have been convinced, but several people in the crowd, recognising the woman's obvious sincerity, had no doubt that she knew something they did not.

'No, hang me instead,' cried another voice.

'Make it me,' roared another.

The crowd began to surge towards the gallows – pushing and jostling each other out of the way in a battle to be first to the noose. In the melee, the woman grabbed her companion and somehow managed to help her escape her guard.

And while the throng fought for the chance to enter

heaven without condition, the two women saddled up their horses and rode away.

* * * * * * * * * * * * * * * * * * * *

Let us leave them here, heading safely into the distance together, while we contemplate one more crucial point that must always be remembered when asking the universe for help.

I am your guardian angel. My job is to help you, without judgement and without question. My values are not your values. My powers are not your powers. My insights are not your insights. But if you trust me and are willing to work with me, I will *always* do my best to get you out of any tricky situation, no matter how hopeless it may seem. And regardless of *how* you got into it, whose fault it was – or whether you 'deserve' that help.

my job is to help you without judgement

I am *your* travelling companion. I am neither male nor female, young nor old ... but I am real, and I do go with you wherever you go.

In that story, one woman reached to her guardian angel in order to help another woman. That's something we have not touched on so far. Shall we talk about it a little now?

CHAPTER 8

On human guardian angels

There are times when you simply can't help other people. There are other times when you can – although it's very debatable whether you should.

If, and when, you ever attempt to play my role, to be a 'guardian angel' to others, you need your relationship with me to be extremely strong and, just as it was in our last story, deeply instinctive.

You really should ask for my help in knowing when and how to offer *your* help – and in recognising when it may be more appropriate to withhold it. Sometimes, people have to find things out the hard way. Sometimes, to protect someone from learning their own lesson is to deprive them of the chance to find independent strength and wisdom.

And then again, there will be times when you can

simply see a genuine way to help someone ... and it's a pleasure and a privilege to be able to extend that assistance. The best way to tell if you're helping in the right way, for the right reason, is to ask yourself what you want in return for your aid. Take it from me, as a full-time guardian angel, there's really only one satisfactory answer to that question: nothing!

Even if you want no more than a little recognition, appreciation, gratitude or acknowledgement, your motives are not entirely selfless. It may still be very enjoyable and gratifying to give your help and that help may well be very effective ... but you need to be very careful. There's a possibility that it may somehow rebound on you – or lead to a future problem either for you, or for the recipient. Now I know that, earlier in this book, I advised you to express and explore gratitude when requesting my assistance. But that's not because I want you to be grateful to me. It's simply because, if you are feeling grateful to the universe for the precious gift of life, you will make wiser choices and you will recognise better options and alternatives. I personally don't want or need your gratitude and nor should you want or need this from anyone you reach out to help.

The real trick is to care deeply enough to make the effort, while at the same time not care. You need to want a good result, yet you need to be open-minded about what a good result actually is.

How can you do this? The secret is to ask for my help in the way you offer yours. If it's your sincere desire to learn how to give entirely selflessly and unconditionally, it's my great pleasure to assist you in that learning process. If you make it your goal in life to be of genuine help to others, you will have the most fulfilling experience of existence that any human being can, will, or ever has had. The wider you cast your net, the more willing you are to reach out, even to people you do not know or have nothing in common with, the greater joy you will experience.

But watch, please, for the dangerous notion that you 'know' what someone else needs, better than they know themselves. Human angels *must* be humble servants and avoid judgement of any kind, or all their good intentions will lead to bitter recriminations. Human angels, however, get a little more freedom and choice than eternal angels like me. I am effectively obliged to assist you if I can, no matter what you ask for. You, though, can stop and say, hang on a moment; I know this person is asking for a particular type of help, and I know that I am perfectly capable of extending it ... but I don't think it would be good for me to actually do so.

Let's take an extreme example. A junkie wants you to give her money so she can buy more heroin. You can spare the money. Her request is sincere. If you were an eternal angel, there would be nothing else to discuss. Rights and wrongs are for humans, not angels.

You, though, *are* a human. Rights and wrongs are crucial to you. And you might see something very wrong with honouring such a request. Or you might want to make it conditional. You might want to say, 'How do you feel about going into rehab and giving up your habit? Would you like me to help you find that kind of help?'

The junkie might be very honest and say, 'No. I just want help getting my next fix.' Or they might think they are more likely to get what they want from you by saying, 'Yes, I'd love that. Please do help. And meanwhile, because I'm desperately caught in an addiction that is agonising unless it is maintained, can you help me pay for another gram of smack just to keep the withdrawal symptoms at bay until I've got a professional to attend to me while I sweat out the living hell of a detoxification process?'

They might sincerely mean this. They might be saying it cynically or, most likely, they might both deeply mean it and hardly mean it at exactly the same moment.

What, under such circumstances, ought you to do? If you want to be a human angel, you surely can't just close your eyes, pretend it isn't happening and walk away. Nor, though, can you decide you are beyond judgement ... and that there are no rights and wrongs.

You can't emulate me in this respect, but you can request my assistance. In deciding to be a human angel you don't sacrifice the right to ask your own guardian angel to take care of you; or, indeed, to help you to take care of others.

While I cannot, as I have explained several times now, ever do anything that conflicts with the agenda of another guardian angel, I can most definitely work in partnership with one of my fellow angels. So there are a couple of things, in a situation like this, that you can profitably request from me.

You can ask me to help you reach your own wisest possible understanding about what's right. I can't just 'tell you' what this is because, as I am a being beyond judgement, I don't actually know. But I most certainly can support you while you think things through, and I can help to bring the right insight, inspiration and example into your life. I can help you to know ... and to be sure that you are not operating out of ignorance, prejudice or fear.

You can also ask me to talk to this person's angel and

see if there's a cause with which we can both co-operate. Then, you may be able to extend an offer to help the person look at alternatives that might be better for them than the thing they think they want. You can then reach your own conclusions about how far you feel it is appropriate to go in offering short-term – as well as long-term – help.

I have picked an extreme example here. So extreme that in talking about a heroin addict, I have used the very impersonal term 'junkie'. I could have said 'a person who is suffering from a big, difficult and debilitating problem'. That person could be a close friend. A friend of a friend. A family member. Ideally, it would not make a difference to you ... for if you really are interested in being a human angel, you will presumably want to treat all humans equally. You won't want to help only the people who live the same sort of lifestyle, or uphold the same kind of beliefs, as you ... or to whom you have been formally introduced! Then again, you'll need to draw a line somewhere – lest you end up feeling responsible for all your billions of fellow humans.

It is, in some ways, a much tougher proposition to be a human angel than an eternal angel, for at least I know where my limits lie. I only have to look after *you*! So, if you are seriously going to attempt to be so wonderfully generous, you'd better refer back to me as often as you can. You had at least better ask me to show you if there *is* anything you can do – and if I can help you to do it without doing yourself damage in the process. You may also want to ask me for guidance in how much to care ... and where to draw a line.

That will be especially relevant if the person who needs help is someone to whom you have a strong emotional attachment. Or if you have a 'parent–child' relationship with them. *All* parents are human angels. They practise this particular form of high spiritual discipline every day of their lives. And every day, too, they must wrestle with difficult choices about how far they should go in imposing their own opinions. Whenever people make their offers of help conditional – or tell others what they may or may not do – they are effectively using their own power to disempower someone else.

Whenever parents help their children to get what they want, they have to weigh up a lot of factors. When should they just grant their offspring's wishes without question? When should they stand firm and refuse? When should they help their child to identify an alternative aspiration? When should their child be allowed to learn the hard way ... or to gain valuable experience from making their own mistakes?

As a human, you live in a world where the quest for the best answers to such questions is endless. As an angel, I live in a world where there's no such thing as a 'best answer'. But of course, here once more I can help *you* to find your own best answer ... if you ask me to.

I can also offer you a word of advice. Unless you actually *are* in a legitimate parent–child relationship with someone, don't ever refuse to grant a request because you don't think that the person who is doing the asking is asking for the right thing. Refuse simply because you feel it would be wrong for *you* to comply. And actually, the same may apply even between parents and children.

One way or another, it's fair to say that to be a good human angel you'll need patience. And wisdom. And kindness. And detachment. You'll need to keep checking yourself for the invisible 'strings' that you may be attaching to your offers or gestures. The knots they tie you up in could take a lot of unravelling.

Don't let that put you off. Do, though, let it make you aware.

Collective ordering

You have to imagine it before you can go ahead and make it happen. Nowadays even footballers are doing this. Business people. Tennis players. You can call it magic, meditation, projection of your goal ... but whatever it is, if it works for them, it can work for all of us. If we all want peace, we just have to project a positive future ...

John Lennon, December 1980 (edited interview transcript)

* * * * * * * * * * * * * * * * * * * *

Is it possible to ask for my help in making the whole world a better place?

You may well think that if millions of people all over the world all wanted the same thing to happen, and they all asked their guardian angels for assistance with this, then the most amazing things could happen. Wars could end. The hungry could be fed. The problems of pollution and global warming could be solved. Couldn't they?

They could indeed – but only through human effort, never through angelic intervention.

Ask me to help you become a more effective force to help the world change for the better and I can most certainly assist. Ask me to help you understand the best causes to support, the most effective organisations to join, and the finest campaigns to lend your energy to and, once more, I can be of use.

Why, though, is there war? Why do some people continue to pursue cold and callous aims and objectives – or entirely selfish, short-sighted goals – when they could be working for universal love, peace and familyhood?

It's got a lot to do with what certain people are asking their angels for. We angels are obliged to honour, without judgement, all requests that are ever made of us. So if one person asks us to help them wage a war and another person asks us to help them stop a war, we're back in the realm of 'cosmic conflict' that I told you about in chapter 1. We angels don't work against each other.

It works in just the same way as it does if you ask the universe to help you get back together with your partner, while your partner is asking the universe for help in putting the relationship behind them. Your respective guardian angels become gridlocked. They both have to sit quietly aside until the pair of you are asking for something that is not mutually incompatible. If, on the other hand, instead of asking me to change the way your ex-partner feels or thinks, you ask me to help you find a way to understand, to accept, and to get along better with them, regardless of how things turn out, there *will* be something I can do.

It's the same with 'great causes'. If five million people are all asking for help in pursuing the same vision of positive change – and just one other person is asking their angel for help in pursuing an agenda that runs contrary to this vision, there's nothing that any of the angels involved can do. We don't run a cosmic democracy. Human beings, however, *do* do this. Movements that gain enough ground can alter opinions and ideas all over the world – often in surprisingly dramatic and sudden ways.

So, if you've got a desire to work for peace, begin by seeking inner peace, forgiveness, understanding, wisdom and tolerance within yourself. Then ask for insight into the best way to help share and spread this. And be open, as you do this, to the possibility that your angel's idea about how you can best serve humanity may not always exactly match that of your favourite cause, political party, or charity campaign. Remember what we said earlier about certainty. It's a very dangerous and divisive thing.

And if you really want to know how guardian angels feel about ending war or creating a world in which the rich share their wealth with the poor, and the powerful use their technologies to reduce pollution rather than create it, I'll tell you. It's all down to consciousness, awareness, sensitivity, discrimination, tolerance and wisdom.

If, or when, enough human beings prioritise these things when they ask their guardian angels for help, those noble goals will be easily fulfilled.

CHAPTER 9

Good ways to get help

Just as we guardian angels alone can't prevent wars and famines, we can't undo what has been done. If something has happened in your life, I can't reverse it so that it never happened. I can't change history. Everything that has ever happened, is happening, and will ever happen, is occurring right now. Even so, we can't turn back time.

I can, however, help you change the way you think about what has happened. I can help you accept. I can help you adjust. I can help you overcome a state of denial or pessimism or deep regret. I can help you see what's good, even in what's bad. I can help you learn. I can help you move on. I can help you broaden your perspective. I can help you compensate. I can help you heal. But only, of course, if you ask me for this.

The trouble is that sometimes when you need my help the most, you find it hardest to ask me for it. Ironically, when you feel down, you lose faith in the notion that anyone or anything can ever get you back 'up'.

Usually, though, before any moment of great fulfilment can occur, there must first be a moment of great emptiness. Just as you have to be hungry before you can really enjoy a meal, you have to recognise your need before you can appreciate the fact that it has been met.

It's true, there are times when you suppress your appetite and don't realise until after you've been fed just how empty you were feeling. If you can 'get away' without an intense experience, you're lucky. But generally the time of triumph and delight is preceded by one of anxiety, uncertainty, or even despair.

It can't be any other way. It's what you have to go through, because it's what, ultimately, gives meaning to the cosmic ordering process and prevents it from being a shallow means of random wish fulfilment.

Still, when you're experiencing need at its most powerful, it's easy to forget that feeling pretty horrible for a while is unavoidable and, ultimately, a sure sign of being on the right track.

Sadly, some people get to this first phase ... and go

no further. They mistake a temporary absence of enthusiasm for a permanent sense of hopelessness. They get overwhelmed by pessimism and assume that they are now suffering from an incurable malaise. They even seek medical treatment for it.

Life has its downs and they are natural. Human beings are not supposed to be happy, smiley people all the time. The world would be entirely full of pathetic plastic platitudes if they were. Sometimes, those downs last longer than you want them to. Sometimes they are inconvenient. But sometimes they exist for a reason.

There may be something you need to face or to understand. It may be unwise or unhealthy for you to rise up out of your dumps until you have dealt with whatever you need to deal with. Chemical sedatives, antidepressants and mood-altering drugs may do people more harm than good if they 'stabilise them' and thus keep them in the midst of a process of transition that they actually need to fight their way through to the end of. It's easier for a doctor to say, 'Here, take this,' than to say, 'Okay, what's really eating you?' Especially when that's likely to require a trip back to early childhood, a careful analysis of an emotional state, a lot of dialogue and therapy ... and more effort than anyone in this busy, modern world has time for.

People suffering from despair, depression, frustration, fear, anxiety, panic and confusion may, in many cases, be simply people who are stuck somewhere in their own cosmic ordering process and who just need to ask their guardian angel for a little more help.

No matter how well we relate, no matter how our relationship deepens and strengthens, there will be times when you find life very, very hard. Remember, please, at such times you're only as young as you feel, you're only as poor as you're pessimistic ... and you're only as helpless as you're proud. Clichés, yes, but true.

Whilst it's important not to become too desperate about anything – or anyone – it's also important to remember that some dreams take a lot of fulfilling. If you want them to come true, you may first have to go through a phase of wanting it very badly and fearing that you're never going to have it.

Help yourself

By now, from all we have said and shared, it must be clear that you can ask me at any time, in any words, in any way. I will always hear you and I will always be willing to help you if I can.

There is, though, an old saying that 'Heaven helps those who help themselves'. Certainly, we guardian angels are well aware that people who believe in the power of their own luck tend to be luckier than those who don't see themselves as lucky. People who work hard, who engineer opportunities for themselves, who believe that it is possible to create their own luck, often manage to go ahead and do just that.

So before asking me for help, it is always a good idea to begin by looking to see what help you can provide for yourself. You may be surprised at how much you can do and how little you require my intervention. The way to find out is always to try. The way to try is always in a spirit of positive expectation ... and it is also a good idea to use some imagination.

You could try drawing two columns and compiling two lists. In one column you could write down all the things that you feel able to do in order to make something come about. In the other column you could ask for the things that you feel your angel could contribute to the situation, for the one piece of magic that you reckon I would have to supply because you couldn't possibly engineer it. In an ideal world, this second column would be nice and short. And your 'what I could do myself' column would be very long.

Through undertaking this exercise, you may sometimes find that you don't actually need anything on my side of the sheet because, on reflection, you feel you can do it all yourself. You may simply need to say, 'Please can I have a little help in actually going ahead and doing all this?' The very act of asking for my help can be enough to cause you to summon a happy coincidence up out of nowhere, to bring forth serendipity, to fill your life with magical, supportive developments that make you feel as though the world really is working miracles on your behalf.

But don't go to the other extreme and decide that you don't want my help or that you don't believe I can help you. Be especially wary of doubt, pessimism, fear, or a general sense of being undeserving. The more you believe that you can't be helped and won't be helped, the more your negative prophesy becomes self-fulfilling, and you end up putting yourself in a position where not only can I not help you, but your friends and loved ones can't help you either, far less yourself!

There will always be some area of your life, no matter how open you may be, how full of hope, how full of light and how full of enthusiasm for the universe, where you feel as though you are stuck in a bad habit or a difficult situation and nothing is going to get you out of it. It's in that area that you most need to seek my assistance because in that area you may be least able to see what you most need to be looking out for.

If you have got a blind spot in your car, you can't drive it safely. You will be fine most of the time, but at some crucial moment, when you have got to make a split-second important decision, you are going to have the wrong information. As we saw in chapter 4, much the same applies when you have a psychological blind spot, when there is something about yourself, your life, who you are and what you are doing that you can't really see. Unless you have got some degree of objectivity, some way of understanding yourself with a fairness and insight that is appropriate and helpful, it is going to trip you up.

Whenever you ask me for something specific, there is a good chance that you are asking for something that is in some way influenced by the area around which you have that blind spot. So, ask me to help you recognise and overcome your own blind spot.

Remember, too, that there is a big, but subtle, difference between asking me for help with something that you can ultimately engineer for yourself – and asking for something that seems entirely beyond your own power to accomplish or achieve. Sometimes the very act of asking me makes it more possible for you to bring about the change by yourself. By asking, with sincerity, you become more open and willing and ready to accept the possibility that something can happen. Whether that then makes it easier for me to work the miracle for you or easier for you to work it for yourself is a moot point.

I'm going to be grateful for whatever unfolds

Yet while remaining open, positive and hopeful, a part of you also has to be open, positive and hopeful about what might happen if that request is not fulfilled. Without surrendering yourself to a feeling of defeat, you must be able to say, 'Okay, I'm asking for this. If I get it, I will be very grateful. If I don't get it, I'm going to be grateful, too. I'm going to be grateful for every moment of my life that I've lived so far. I'm going to be grateful for every experience I'm currently having, even if it isn't the experience that I think I ought to be having or wish I was having. I'm going to be grateful for whatever unfolds.'

Choice

Asking for my help is to a great extent about making a choice. It's about saying, 'Yes. This is what I want.' But, at the same time, I urge you to remember that whenever you make a choice, you actually reduce the opportunities that might otherwise be available to you.

That's why, as I have said before, it is always wiser to request general rather than specific help. The more particular and precise you are about what you are requesting, the more likely you are to end up with exactly what it is that you have asked for, consequences and all, drawbacks and all, unexpected and unforeseen side effects and all.

In a moment, I am going to talk you through a little process that might help concentrate your mind and make sure that you are clear, as you communicate with me, about what it is you want me to do for you. It's a powerful technique that I am about to reveal to you, so first of all I feel a duty to reiterate a couple of points – and to explain one more thing.

Every time you turn on the TV you are told what you ought to be buying, how you ought to be leading your life, what kind of house you ought to be living in, what kind of car you ought to be driving, what kind of food you ought to be eating, and what kind of relationships you ought to be having. It's all described for you in the movies, in the books you read, the magazines that you leaf through and, of course, it's also described in your own fantasies.

You dream about the situations, the possessions, the developments that will do the trick for you, that will turn your life around, that will take you from sadness to happiness, fill you up, make you content and bring that glorious fairy-tale 'happy ever after'. Only the reality doesn't work that way.

Think back in your life over the various things you have aspired to and attained. Some have been wonderful and others have not been so ideal.

Sometimes when people have backed themselves into a corner, they don't admit they've made a mistake. They say, 'What a great corner this is. What a great problem I've created for myself. What a great difficulty I now have. I'm so glad I've got it.' Then they start listing all the advantages. This could be seen as looking on the bright side, but it could also be seen as justifying a mistake and compounding an error by refusing to concede that it exists!

That's the drawback with asking me directly and clearly for what you want. You don't really know what you want. Nobody really does. You know what you imagine you want, but you are basing your imagination on your experience, and your experience is woefully limited.

Think of yourself as a small child, being taken out for a meal by a kind, indulgent parent. The cosmos wants you to go to the finest restaurant where the most wonderful food is being cooked by the most talented chefs to the highest possible standards. There is a great range and variety of glorious dishes from all over the world. They are all available for your delectation, all at a price that you can easily afford and, what's more, there is a tasting menu. You can experiment with as many different offerings as you care to. There is no problem if you don't like something – it's fine, you can move on to the next dish.

But since you are a child, all this may escape you. You may just say, 'I want a burger.' Or, 'I want beans and chips.' Or, 'I want a chocolate milkshake. I want what I know, I know what I like, I'm nervous of anything else and I'm against it almost on principle.'

Maybe that's fine if it will satisfy you. But if you do settle for the routine, without experimentation or exercising your imagination, you might never know what other things you could have learned to love.

However, you are not a child. You are an adult who has come to this earth specifically for the purpose of exploring and discovering all the deep wonder and magic that is available. To have prejudices against some of it is very limiting. Ultimately, it erodes your ability to be happy. The more you feel that you can only be happy if you have a particular sort of experience, the less you can do and the less fun and adventure you can have.

Now, it is also true that some foods are an acquired

taste. But the foods that you learn to like can sometimes be more pleasurable than the ones you form an affinity with the moment they touch your taste buds. Asking me to choose for you, at least sometimes, is like going into a restaurant and saying to the waiter, 'Go ahead and bring me the best dish you have got and, when it comes, if I don't like it, tell me what it is I ought to be liking about it. Show me how to appreciate it.'

show me how to appreciate it

If, despite the waiter's best efforts to increase your experiences, you still feel that this dish is not for you – well, in this particular restaurant, you can send it back and request something else, no problem. You will at least have known more than you did before.

To show you how to get my best attention at all times, in all ways, I now have a request to make of you. When you're visiting my restaurant and I am being your waiter, at least ask me to show you the entire menu of choices before you order!

My secret agenda

Is that all I am – a waiter in the restaurant of the universe?

Well, yes and no. I am the link between the transient, ever-changing, always striving, physical you, and the perfect, permanent, ecstatic force of life that ultimately allows you to be all that you are. My ultimate purpose is to help you rediscover the lost memory of your life before you came to this world and to function, while you are here, as a person full of inspiration and appreciation.

I can help you get anything in this world that you want and you can use me to achieve this at any level that pleases you. I will bring you cars and houses, lovers and loved ones, career successes and social triumphs. If you want these things, you just have to ask me and, if I possibly can, I will help you get them.

But if it ever occurs to you to pose a question to me, momentarily, in between asking for what you want, here is what I wish you would ask.

I'd like you to ask me what *I* want. Because what I want ... is for you to want what you need! And, if you don't know the difference between what you want and what you need, I want you to know that difference!

I want you to be willing to be guided, willing to

learn, willing to trust, and willing to discover. Then, instead of performing trivial tasks that may or may not bring you any real lasting satisfaction, I can set about the business of bringing you happiness every single day. It's not a case of either/or. It's a question of *and*. Ask for insight and wisdom, then when you get it you will know what else you really ought to be asking for.

CHAPTER 10

Requesting angelic intervention

Ultimately, it's very important that you reach out to me in the way that feels natural and right to you. There are no hard and fast rules about how we have to communicate. Or when. Or where.

But as you also know, there are some times that are better than others. And you can always tell, in your heart of hearts, whether you have arrived at such a time.

So let's assume, shall we, that this is the moment.

If it isn't, put the book down and come back when it is.

Now.

Where are you?

Is it quiet? Is it comfortable? Is it peaceful?

Are you on your own? Do you have plenty of

time? Can you spare a whole hour if you need to?

Good.

Cosmic ordering *can* be done in a hurry ... You can reach out to me anywhere you are, any time you need to. But there are some processes in life that you really don't want to rush, especially when you have a profound request ... or a deep desire to move your life on to a new and happier level.

So.

Have you got an incense stick you can light?

Or a bottle of pure, exotic aromatherapy oil?

How about a candle? Can we draw the curtains and lower the lighting a little?

You don't actually *need* any of these, any more than you need to be alone. You certainly don't want to become reliant on them while you are talking to me. I'd hate you to feel that I will only listen to you if you create a romantic atmosphere. But at the same time, if you want to enter a state of great concentration and really deepen your sense of connection, a few touches like this can be very helpful.

Er ... this is a silly question but ... you *are* sober, aren't you?

If you've had a drink, a joint, a pick-me-up, a wind-me-down, or a spin-me-out ... you might want to wait until you're no longer under a chemical influence. It's not essential, I'll accept your request no matter what frame of mind you are in when you make it, but you'll like the experience a lot better if you're as calm and as close to a natural state of being as you can be when we speak.

Right.

Let's do a little breathing.

Draw one in. Hold it just a moment.

Savour the air in your lungs.

And now slowly expel it.

Repeat the exercise.

As you do, consider this.

Where is that air coming from?

What, within you, keeps drawing it in ... even when you're not making any effort to make the process happen?

Somewhere within you, there's a power that's greater than you are.

It runs systems that lie *behind* the systems you have conscious control over.

It is not, I promise you, an automatic, 'unconscious' mechanism.

It's a function of your own higher consciousness.

It's evidence, indeed, of your higher self in action.

Let's show it some respect and join with it in the taking of a few more breaths.

Put the book down for a moment and just breathe.

Watch those breaths come in. Watch them go out.

Give thanks to the part of yourself that tirelessly watches them come in and go out, hour after hour, day after day . . .

When you're ready, pick up the book again.

Here's a further question to ponder.

What's the difference between the mechanism in you that governs all that inhalation and exhalation, and the mechanism in every other human being on this planet?

Thank you.

That's the point.

There *is* no difference.

People's higher selves are as similar, in purpose and consciousness, as their 'lower selves' are seemingly different.

Okay. Keep noticing those precious breaths coming in and out. Take them slowly and gently.

And let's ask another question.

What's the difference between the force that drives your breath and the force that keeps this planet spinning on its axis as it hurtles endlessly round the sun?

Thank you again.

Now we're getting somewhere.

There's a very *big* force inside you.

And ultimately, it's *not* inside you.

You are inside it.

Let's keep that breathing going.

It's time to consider something else.

Think, please, of a tree.

A nice big one, with a tall trunk, high branches and many leaves.

Imagine, for a moment, that you are just one leaf on that tree – a little leaf, but one that wants to be as green, as shiny, as healthy and as 'in the right place' as it can possibly be.

Are you on your own in this endeavour?

No.

You have the support of the entire tree.

The other leaves around you may be too busy fulfilling their own potential to give you much more than a little 'social engagement', but the branch wants the best for you. And the trunk is right behind it. And so, too, are the roots.

Whatever you need to be the best that you can be, they'll provide, if they possibly can.

Okay. Keep breathing.

You're not a leaf. Nor are you a tree.

But we're getting somewhere here.

Somewhere important.

Take another breath and take, in your mind's eye, another look at that tree. Think of the roots, going down, deep into the ground.

Are they a part of the earth or are they separate from it?

Forget whatever you learned in biology or horticulture class. There's a technical distinction between the end of the root and the beginning of the earth ... but ultimately, they are as one – as are all things in, and on, this earth.

Just take a little leap of creative imagination and cosmic visualisation ... and another breath.

Another breath. A deep one, please.

I want to take you down into the earth.

And I want you to ask yourself one of the most important questions any human being can ask.

What *is* this planet on which you live?

Are you separate from it?

Or are you a part of it?

If you are a part of it ...

... perhaps you ought to think of the whole earth as one living organism. Just like that tree with its branches and its leaves.

The leaves may think of themselves as 'independent entities', but they're all joined to the rest of the plant.

The people of this earth may see themselves as separate to it and to each other, but they are physically connected to the planet and thus ... to all other creatures upon it.

Breathe again. And consider this.

The earth is but one tiny ball of rock in a cosmos of infinite size.

It is, however, as much a part of that cosmos as any other planet, moon, comet or star.

And so just as you can think of this earth as one big living organism, you can think of the cosmos in the same way.

If the entire universe is *one* big thing, who is to say where the most (or least) important part of it is?

Another breath.

Another, too.

And now, although I know we've just spent a long time in communication, I'd like to introduce myself to you, formally.

I am you. You are me. We are each other. I am a part of you that was never born and thus can never die. I am a part of you that has existed forever and therefore always exists. I am what some people call your higher self. Others might say I am your true self. I am the divine energy that pulses through your being with your every breath. I am not your personality, your body, or even your heart. I am your consciousness. I am the part of you that is 'aware of itself'. I am also the part of you that is aware of its connection to every other living thing on this planet and, indeed, in this universe.

Breathe, please.

And again.

Be still. Be quiet. Be calm. Be trusting.

Now, let's contemplate another question.

Why are human beings supposedly 'superior' to other living beings?

Some say it's because of their capacity to be aware of their own existence.

Right now, you're breathing.

You're aware of your own existence.

You're also aware of your ultimate interconnectedness with everything.

So now, consider, one last time, please, the power you are about to wield, the strength you are about to summon, the request you are about to make of me, your guardian angel, your higher self.

Is there anything in it that could cause harm to others?

Is there anything about it that's hurtful or vengeful?

Is there anything about it that's petty or spiteful?

Is there any way you might modify it so that it's as

'pure' and 'noble' as it can be? Remember, you don't want to put me into a cosmic conflict.

Even at this last minute, it's not too late to rephrase it if inspiration suddenly strikes you.

Okay.

When you're comfortable, draw another breath and see before you, please, the objective you want to attain.

Envisage it.

Embrace it.

Own it.

Be it.

Believe it.

Live it.

And ...

now ...

let it go.

This, if you want it to be, is the moment when you can write your request on a piece of paper and burn it in the candle flame.

Otherwise, just gently allow that image in your mind's eye to evaporate.

Don't turn it into disappointment. Or emptiness. Or regret.

Turn it into love. And trust. And inspiration.

You may find this comes extremely naturally to you.

If not, here's how to do it.

First, keep breathing.

And briefly shut your eyes (when you need to refer to the book again, you can open them, of course).

Now, envisage yourself a few years in the future.

Strong, fulfilled and deeply happy.

You have everything you want.

However, right now, you don't necessarily have what you want.

Either you had it and you realised later that you didn't need it ...

... or you never got it.

Why would you care?

You're completely content.

Radiant with bliss and joy.

All you have to do is think of yourself feeling good.

Feeling really, really good.

Remember: when you feel good, you just don't care about why you feel good.

All you know is that you've got that feeling.

It's important to see yourself in that state ...

Because, ultimately, that's the state you want to order up for yourself, regardless of what you are asking me for.

Breathe again.

And again.

Envisage yourself as bigger than the world.

Try imagining that you're some giant, ethereal figure so large that you can hold the whole earth in the palm of your hand. Imagine this planet somehow sitting in one of your hands.

Allow yourself to be godlike.

See yourself, your true self, your highest self, as the most powerful thing in the whole of Creation.

Because that's what you are a part of.

So, ultimately, that's what you *are*.

Now, reach out with your mind to this enormous, kind, intergalactic-sized, cosmos-spanning version of yourself and say ...

'Help me, please.'

You're no longer asking some vague disembodied force called 'the universe'.

You're not appealing fearfully to some disapproving authority figure of a god or goddess.

You're not even asking me.

You are asking yourself, in your highest form of being, for a little more help in remembering who and what you really are ... and a little more help in achieving what you truly need.

Breathe again.

And again.

Don't worry about whether the way you are doing any of this is 'right' or 'wrong'.

Don't be concerned if you can only 'half imagine' any of the above.

If you can see it at all, no matter how dimly, you're seeing it.

And feeling it.

And doing it.

Now, it's time to transcend the laws of time and space.

Or at least, the conventional ones.

You see, the part of you that's as big as the entire universe is not trapped by the clock and calendar, as we understand it.

Nor, come to that, is it bound by laws of birth and death ... though that's another topic for another time.

One thing's for sure, though.

It knows the future.

Just as surely as it knows the past.

Breathe again.

Ask yourself, your higher self, your wiser self, your future self, what you ought to be asking for?

Ask for the wisdom to recognise your own higher wisdom when you hear it.

Ask, too, for the confidence to let your own higher wisdom intercede.

And then breathe again.

Stay in your comfortable position for as long as you want to.

Fall asleep if you can.

Your order has been placed.

Your request has been heard.

Feel free to think about it as much as you like.

Or never to think about it again.

It makes no difference.

You're going to get the closest thing to the fulfilment of your request that the cosmos can possibly provide.

And the cosmos is a pretty bountiful place.

It may take a week or two to come.

It may take a year or two.

But come it will.

Remember, you absolutely *don't* have to do this every time you want to connect with me, or ask the universe for something ... or place a cosmic order. There will be many a time when it all happens naturally in a single thought or a single breath. But if ever you're lacking clarity, or needing focus, or wanting to summon a lot of positive change into your life ... this way will work for you.

Afterword

Once upon a time, there was a guardian angel, who had spent many long years looking after a very special woman. This woman had worked hard to make the most of her time on planet earth. She had tried to be kind, wise and generous. She had kept an open mind wherever possible and had done her best to help make the world a better place. Not just for her friends and family, but for all her fellow beings. Still, though, she had had her share of hard times and struggles. She had known sadness, loneliness, fear and need. Just as she had also known adventure, excitement, satisfaction and success.

Her guardian angel had been with her through it all, watching her, helping her, whispering to her and even, in some instances, extending the kind of assistance that blurred the angel's professional boundaries. Once or twice the angel had taken to jumping up and down excitedly in front of the woman, urging her to seize the opportunities that she was apparently oblivious to. Of course, the woman had never noticed. And the angel, naturally, had not minded, being blessed, like all angels, with infinite patience and understanding.

At the end of a long and fruitful life, the woman left her body and finally she became able to speak to her angel face to face. Together, they floated on the special white fluffy cloud that the universe supplies at such a time of transition. They sat down to share the in-depth review of her life that the woman was about to watch. After studying the various recordings from numerous angles, the woman learned much about what had really been going on, unbeknownst to her, at all the key moments of her existence. Finally, the woman and the angel reached a point where it became appropriate to review their own interaction.

They found themselves looking down on a diagrammatic representation of the woman's life. Her timeline was spread out before them as a series of footsteps on a long, sandy beach. To the angel's surprise, for great stretches at a time there was only one set of prints. 'I was with you all this time,' said the angel. 'I recall quite distinctly that these were some of your most joyous, meaningful moments. Where then are my footprints in this picture? Why are they not right by the side of yours?'

The woman looked at the angel, slightly surprised. 'I know that I've only just entered the eternal realm,' she said, 'but

I reckon I can answer that one for you. You weren't walking by my side; you were sitting on my shoulder. I felt you with me everywhere I went and, instinctively, I clung to you so tightly that I never put you down. What confuses me are those points over there, where we can definitely see my footprints and yours together. These were some of my toughest moments. I went through a lot of stress and hardship and I would have thought that there should have been only one set of footprints then. Yours! *Surely, that was the point when you were carrying* me, *wasn't it?'*

'Ah,' said the angel, 'that was where I was trying to carry you, but you wouldn't let me. I was running along, trying to keep up with you, shouting, "Please let me help." But you kept marching along alone, insisting that you had everything under control and that you knew how to cope.'

'No, I didn't,' she said. 'I distinctly remember, I was constantly requesting your assistance.'

'Yes,' said the angel, 'but you kept asking me for things that I simply couldn't get for you and ignoring me when I offered you options that would have made a world of difference, though not quite in a way that you might have thought you wanted them to.'

They sat for a while longer in silence, surveying the scene. Finally, they turned to each other and said, simultaneously, 'Still, we didn't do badly, did we?'

And they both smiled.